*Advanc*

"Should atheists and scientists—especially scientists who are atheists—just be disdainful of religious people, however thoughtful the believers may be? In *Sharing Reality*, Jeff T. Haley and Dale McGowan not only insist that disdain is not the way—they lay out in rich detail a better way. With wit and a compelling command of the facts, the authors are persuasive that their answer is correct—and vitally important."

—Ed Buckner, former president of American Atheists and author of *In Freedom We Trust: An Atheist Guide to Religious Liberty*

"This is a clear, candid, and iconoclastic contribution to the contemporary God debates. Haley and McGowan challenge believers and atheists alike with their radical vision of religion's future."

—Russell Blackford, Laureate of the International Academy of Humanism and author of *Freedom of Religion and the Secular State*

"*Sharing Reality* is an important and very readable work for the secular movement in world politics, which is becoming increasingly infected with the excesses that traditional religion invites. In *Sharing Reality*, the authors make a heroically diplomatic plea for secularism that would unite people of all stripes, including religious fundamentalists."

—Shadia Drury, Canada Research Chair in Social Justice at the University of Regina

"This is the best case yet made for a different approach toward secularizing the world by working with religious populations to liberalize their faith. Anyone who would balk at such an approach simply has to contend with the arguments in this book."

—Richard Carrier, author of *Sense and Goodness without God, On the Historicity of Jesus*, and *Proving History*

"We have long been in need of this book. In our often-stonewalled conversation with religious believers, Haley and McGowan bring a vital contribution for the dialogue of both coffee tables and conference halls. Packed full of insights, *Sharing Reality* turns our focus toward the critical component of science acceptance and exposes how the language we often use hinders real progress."

—Drew Bekius, former pastor and president of The Clergy Project

# SHARING REALITY

# SHARING REALITY

*How to Bring Secularism and Science
to an Evolving Religious World*

BY JEFF T. HALEY
AND DALE McGOWAN

PITCHSTONE PUBLISHING
Durham, North Carolina

Pitchstone Publishing
Durham, North Carolina
www.pitchstonepublishing.com

10 9 8 7 6 5 4 3 2 1

**Library of Congress Cataloging-in-Publication Data**

Names: Haley, Jeff T., author.
Title: Sharing reality : how to bring secularism and science to an evolving
 religious world / by Jeff T. Haley and Dale McGowan.
Description: Durham, North Carolina : Pitchstone Publishing, 2017. | Includes
 bibliographical references.
Identifiers: LCCN 2017024473| ISBN 9781634311267 (pbk. : alk. paper) | ISBN
 9781634311281 (epdf) | ISBN 9781634311298 (mobi)
Subjects: LCSH: Religion and science. | Science. | Secularism. | Religion.
Classification: LCC BL240.3 .H356 2017 | DDC 201/.65—dc23
LC record available at https://lccn.loc.gov/2017024473

# CONTENTS

# 1

# A FRAMEWORK
# FOR SHARING REALITY

In November 2006, a science advocacy group called The Science Network gathered luminaries in science, philosophy, and the humanities for the first Beyond Belief symposium at the Salk Institute for Biological Studies in La Jolla, California. The event was a response to recent popular efforts to reconcile the tenets of science and religion so that they might be viewed as equally valid explanatory systems, with the result of blurring the distinctions between the two and obscuring the essence of science. Among the presenters was Richard Dawkins, a celebrated evolutionary biologist and easily the most famous advocate of antitheism on Earth. His talk was vintage Dawkins—brilliant and accurate in content, acerbic in tone.

Also attending the event was the astrophysicist Neil deGrasse Tyson, who asked for the microphone after Dawkins' talk. After acknowledging that Dawkins' words had come out beautifully and articulately "as they always do," Tyson added, "Let me just say, your commentary had a sharpness of teeth that I had not even projected for you. So I felt you more than I heard you.

"You are Professor for the Public Understanding of Science," Tyson continued, leaning on the word understanding, "not professor of delivering truth to the public. And these are two different exercises." "Being an educator is not only getting the truth right," he argued. "There's got to be an act of persuasion as well. Persuasion isn't always, 'Here are the facts—you're an idiot or you are not,' but, 'Here are the facts and here is a sensitivity to your state of mind.' And it's the facts plus the sensitivity [that], when convolved together, create impact. I worry that your methods, and how articulately barbed you can be, end up simply being ineffective."

It's significant that Tyson didn't complain that Dawkins' approach was unpleasant or disrespectful. He said it was ineffective. His argument is that Dawkins' own presumed goal of convincing others that his ideas are worthy and important is short-circuited by a failure to consider the state of the mind on the receiving end of those ideas.

The conflict between religion and science is often described in two very different ways. Some believers insist that the conflict is nonexistent—that religion and science can and do coexist perfectly, often by claiming that religion and science do not overlap. At the other extreme are some of the so-called New Atheists who consider the conflict both real and utterly insoluble by any means short of the end of religion.

This book focuses on a third path, one that acknowledges the problem and offers a solution that doesn't require the (highly unlikely) dismantling of religion. Instead, it relies on a practice that already exists—the ability of religions to adapt—to evolve their positions in light of new knowledge.[1] Atheists who scoff at such an idea have failed to recognize the profound changes in many religious expressions in recent decades. This includes both a liberalizing of social positions and a growing awareness of the importance of embracing scientific truth. The best way forward is by promoting and encouraging the expansion of this crucial process within contemporary religion—a process that happens first for individual members of religions before it is embraced by religious leaders.

Humans are social animals and they will always form social groups, some of which will continue to call themselves "religious."

The prediction made over a century ago that religion would soon fade away has turned out to have been wrong.[2] However, religions will change and adapt to avoid conflict with science and secularism. The open question is how long this process will take.

*Sharing Reality* is a book with two purposes: to convince the reader that accepting science and secularism is essential to building a better world, and to help the reader in turn convince others of the same. If anything, the authors' task is easier than yours. The fact that you've picked up this book means your mind is likely receptive to the values of science and secularism already. At the very least, you are open to being convinced. But if you are like most people, your family, your workplace, and your Facebook feed all include many people who are not so inclined. For you, more than for us, there is a necessary act of persuasion involved in promoting these vital ideas. So, in addition to building the case, we will examine time-tested ways to cultivate a receptive mind in your listener.

We will explain how those with religious beliefs can further evolve their thinking to be consistent with science while retaining some of their religious beliefs about values, and why it is important for them to do so. We will look at how much humans have progressed through cultural evolution and make specific suggestions for furthering cultural evolution through the spread of secularism and science, from one person to another and through media and culture, without ending religion. The overarching message is that an accurate understanding of reality can make an important difference for each person, for societies, and for all humanity.

Because religions offer many benefits to their adherents, they cannot be eliminated. Some New Atheists ask everyone to leave their religions. This is unachievable and unnecessary. We should help all religions evolve to be consistent with a modern understanding of both political and scientific reality by directing our sharing of secularism and science to both religious people and their leaders. We should change the definition of *religion* so that any group that accepts all the conclusions of science can also call itself "religious," even if the group's beliefs are atheistic.

Finally, we argue that it is not enough to be a skeptic and a critical thinker and a secularist and an agnostic. To make progress in our cultural evolution, it is important that each person accept the scientific consensus on all topics, including such topics as homeopathy, astrology, fate, karma, unlucky days or numbers, and safety of vaccines and fluoridated water supplies. Accepting science on these topics is more important than becoming an agnostic or atheist. Once people accept the evidentiary basis of science, they usually eventually give up their god beliefs. The most important cognitive transition that a majority must achieve to move cultural evolution forward is not from theism to atheism; it is from using only traditional ways of knowing to also using every day the scientific way of knowing.

In chapters 3 and 4, we offer twenty-five specific suggestions for how you can help your friends, family, religious leaders, and the entire world learn to accept secularism and science to make further progress with their cultural evolution. Some of these suggestions are old, but most of them are new. (See the appendix for the full list of suggestions.)

## Beyond the Tribe

Tribalism—a strong loyalty to one's own group and hostility to other groups—is as old as humankind and as young as the latest Super Bowl. A tendency toward tribalism was wired into us by 120,000 generations living in the Paleolithic era.

For two million years, the genus *Homo* was poised on the edge of extinction, living in small bands competing for scarce resources. Cooperation within a small group was adaptive, but cooperating with another group passing through your territory was not. You wouldn't benefit from doubling the number of mouths to feed while available food in the area stayed the same. Those who distrusted people who dressed, looked, talked, or acted differently were more likely to survive. Nationalism, militarism, racism, and the fear of immigrants and all things foreign are natural modern descendants of what was once a very effective survival strategy.

But in a densely populated world, one in which people of different races and religions must live in close proximity with one another, natural tendencies to fear and distrust those who are different can be dangerous for both the fearful and the targets of their fear. As ideas of civilization have spread, they have reduced tribal conflict, something that has benefited everyone.[3] This spread of ideas that make life better for all is an ongoing aspect of human evolution. And now, societal norms and ideas are changing at a very swift pace due to electronic communications.

In the last two thousand years, a very short span in human history, widespread belief in a few dominant religions came to replace local beliefs, spreading to engulf people of many tribes and creating a kind of super-tribal identity. What was once merely tribal conflict became religious conflict, and the increase in scale and creedal loyalty often came with an increase in violence and intolerance. Since the Enlightenment of the seventeenth and eighteenth centuries, influenced by new discoveries in science, increased confidence in reason, and development of political philosophy, religious conflict has lessened in much of the world. Though many people have remained at least nominally religious, religions in many places have evolved to accept principles of tolerance to reduce conflict and violence.

But in other parts of the world, particularly parts of the Middle East and Africa, religions continue to lag in adoption of these ideas of tolerance and coexistence. Increasing globalization has now brought violence driven by differing tribal and religious views from the Middle East to other parts of the world.[4] Unfortunately, such practices as the abduction of women and murder by members of the Islamic State, Boko Haram, and al-Shabaab are natural, tribalistic human behaviors. Such natural tendencies can only be overcome by further cultural evolution.

But even where religions have generally become more tolerant and ceased inciting violence against others, they typically continue to cause family discord, social discord, and oppression of those who do not follow their precepts. For example, religious or similar beliefs that deny the use of modern healthcare cause suffering and death as people refuse to accept scientific knowledge. Refusal to accept the truth of facts as revealed by science hobbles the mental abilities and physical well-being of individuals all over the globe.

## WHY PROMOTE SECULARISM?

How can we as individuals help all cultures evolve to reduce tribal and religious conflict and reap the benefits of science? How can we help our own religions continue to become more tolerant? How can we help reduce conflict and suffering from religion, false beliefs, and science denial among our family, friends, and communities? One powerful answer is to promote secularism.

Unfortunately, not everyone agrees on the meaning of the word *secular*. For most of the discussion in this book, such distinctions don't really matter, but it may be useful to know that there are at least three different meanings in current use. Let us call them the general meaning, the core-values meaning, and the nonreligious meaning, a newer, more narrow meaning that implies a rejection of religious belief and organizations.[5] Because secularism is one of the key concepts in this book, we must take a moment to clarify which versions of secularism we are defending.

### THE GENERAL MEANING OF SECULAR

The general meaning is derived from the Latin word *saeculum,* meaning "of this world." Secular thought in this respect is centered on the concerns of the natural universe around us, whether or not supernatural entities or forces exist.

### THE CORE-VALUES MEANING OF SECULAR

The core values meaning of *secular* involves three core values[6]:

1. **Equality.** This value requires that people of different tribes, religions, and beliefs are equal before the law (i.e., in all aspects of government).[7] It requires freedom of speech on all topics, including facts, values, and religion (no apostasy, heresy, or blasphemy laws).

2. **Liberty.** Freedom from coercion to take actions that violate one's conscience unless such coercion is required to avoid harming others; freedom of speech, of belief, and of privacy. In a secular

society, no religion or tribe is allowed to oppress someone who doesn't voluntarily subscribe to that religion or tribe. All people are free to practice their religion as they wish, or none at all, provided they do not endanger others or suppress their freedoms.[8] Both the first and the second value require freedom of speech.

3. **Truth in Government.** Government decisions should be based on objective facts that nearly everyone accepts as true. This value holds that no facts inconsistent with science asserted by a religious/tribal/ethnic group may be endorsed by government. When participating in public discourse to decide on public policy, arguments should not be based on asserted facts that only some members of the society believe—they should be based only on facts for which there is a broadly shared consensus by people of all religions and people of no religion.[9]

The core values of secularism are mostly about tolerance of other people and their values, but secularism is intolerant of people not accepting these core values. Rejecting these core values tends to make society more violent and intolerant. When such a rejection of tolerant values leads to action, as it often does, it cannot be passively accepted by those who value a peaceful and tolerant society.

Many religions have evolved over time from cultural hegemons to willing members of a diverse cultural landscape. Some members of all religions, including Catholicism and Islam, support secularism according to the core-values meaning.[10] The Southern Baptist denomination even includes church-state separation directly in its central doctrinal statement, the Baptist Faith and Message.[11] That doesn't mean that Baptists think the other members of the religious landscape have their facts right. Secularism doesn't require that. It simply sees the mutual benefit of granting everyone, to put it simply, the right to be wrong. People who use the word secular according to the core-values meaning might claim to be a "secular" Christian or a "secular" Muslim or a "secular" Jew: they adhere to the teachings of their religion but they do not try to foist these views on others or on government, and their arguments for public policy are based on facts that nearly all people agree on.

In contrast, many religions often claim that governments and laws must be based on their views on facts or values, even when those views conflict with those of other religions. Secularism upholds and defends the right to hold such views themselves while firmly declining to uphold or defend their exclusive claim on government or culture. Secularism requires that religions be tolerant of other views and demand no actions by government or any institution that would oppress people of other religions. It's a challenging balance, but one well worth striking for the collective good.

The core-values meaning of *secularism,* which requires evolution by most religions but is not incompatible with the more evolved religions, is essential to minimizing conflict in the world. There is no other practical option.[12]

### THE "NONRELIGIOUS" MEANING OF SECULAR

More recently, many authors and activists have begun to use the word *secular* with a third meaning that encompasses the core values of tolerance and coexistence described above and also requires more of people to be labeled as *secular.* These authors use the word *secular* to refer only to people who do not participate in any religion and are not active believers in any religious doctrine, including people who just ignore debates about religious belief or may not have decided whether they believe in anything supernatural.[13] In this usage, you cannot be a *secular* Christian or a *secular* Muslim or a *secular* Jew.

This third meaning of *secular* as "nonreligious" is narrower in that it describes fewer people. It is less helpful for reducing conflict; it robs secularism of its power to heal divisions and advance the best common interest of humanity. This meaning is synonymous with "nonreligious," and this makes the meaning inherently unclear because, as we discuss later, "religious" has no clear meaning. The diversity of groups that consider themselves "religious" is broad and becoming broader, which gives the word "nonreligious" less and less definition as we move forward.

## THE END OF SECULARISM:
## A CAUTIONARY TALE

*What follows is a brief parable to illustrate why secularism is a value that benefits everyone, both religious and nonreligious, both church and state.*

It was a cold, bright February day at the White House. The recently inaugurated president of the United States strode to a lectern in the Rose Garden where members of the religious press had assembled for an announcement.

The details in the press release had been few and vague. An executive order would be announced, one that fulfilled a campaign promise made by the conservative president.

The president cleared his throat. "Thank you for being here. I have invited you here today, as representatives of our nation's online, print, and broadcast religious media, to witness a longstanding dream of the American people coming to fruition." He scanned the small crowd. "Despite attempts to deny our religious identity, you know that ours is a faithful nation, a nation of believers."

All heads nodded as one, with smiles all around.

"In light of this faithful heritage, it is my privilege to announce that as of noon Eastern Time today, Executive Order 14103 takes effect, ending the separation of church and state in the United States of America."

The press buzzed, some elated, others confused. Cameras whirred and clicked.

"I'll take a few questions. Yes, Steve from the *Evangelical Times.*"

"Is it fair to say that this at last affirms that ours is a Christian nation?"

"That's correct. This order simply affirms what most of us have always known to be true: that the United States of America is a Christian nation. And when the House reconvenes next week, I expect its first order of business will be giving this affirmation the force of law."

At the mention of "Christian nation," the smiles dropped from 29 percent of the faces. The Baptists, however, beamed.

"Are you confident in swift passage of the law?" asked Margaret Maples from the *Baptist Press*.

"Completely confident. Oh, there will be some strangled cries from the usual suspects, no doubt," he said, drawing a ripple of laughter. "But as of the current session, my party controls both chambers of Congress. Even if that weren't the case, it's unlikely that more than a few socialists would ever oppose something with such deep and wide support among the American people. Yes, Melissa."

"Are you at all concerned about backlash from minority religions?" She smiled nervously at the reporter to her right, in beard and yarmulke, who simply sighed.

"By my reading of the Constitution," continued the president, "this is a country that runs on majority rule. But ours is a religion of generosity and peace, so all Americans can expect fair treatment regardless of their own beliefs."

Another reporter slowly raised his hand.

"Yes, John from the *Wesleyan*."

"You say it will be a Christian nation. What variety of Christianity? There is no generic version, of course. Every expression has its own traditions and beliefs, so . . ."

"Exactly right," said the president. "So once again, we turn to democratic principles of majority rule . . . or barring that, plurality. The largest Christian denomination will now be our state religion."

"And that would be . . . ?"

"Catholicism."

There was an audible gasp, followed by silence. The smiles had now vanished from 76 percent of the assembled. The Baptists looked incredulous.

"Over half the country is Protestant!" someone shouted.

"True. But Protestantism is not a denomination, and to paraphrase John, there is no generic Protestantism that could serve. The social doctrines of the Pentecostals and the United Church of Christ, for example, are nearly opposite each other. Southern Baptists are the largest Protestant denomination, it's true, but at 11 percent, they are well below the 24 percent represented by U.S. Catholics."

All eyes turned to the journalists for *Osservatore Romano* and *Catholic Reporter*, who shrugged and smiled.

"So . . . what will this look like, exactly?" asked Julie Winthrop of Religion News Service. "I mean, in a practical sense?"

"We expect the rollout to take about two years," the president replied. "To begin, all non-Catholic military chaplains will be replaced with Catholic priests. Each session of Congress will begin with a Catholic prayer, invoking the protection of Mother Mary and the Holy Father in Rome. All legislation will be reviewed by a council of cardinals to ensure conformity with Catholic doctrine. Converting our public schools to Catholic schools may take a bit longer, but we expect to have catechism classes in the curriculum by next year. Revisions to sex education will follow soon thereafter. Prayer will return to our schools, something that your media have long advocated. Each prayer will of course include a petition to Mother Mary and a blessing on the Holy Father."

"My children are not going to a Catholic school!" fumed a correspondent from the Mormon *Ensign*.

"You will still have the option of a private education in the faith of your choice."

"But my taxes support the public school system! I should be able to make use of those schools without putting my children in the hands of Papists!"

The president looked at him sternly. "I would be very careful about using such language from now on, Joseph. Now returning to Julie's question: the executive order will render birth control, abortion, and gay marriage immediately illegal."

"And the death penalty?" asked the journalist from *Catholic Reporter*. "I assume that will also be illegal."

"Well, there has to be some give and take. I know opposition to capital punishment is an important social position for Catholics, but this is a question of state's rights. So instead, we will need the Catholic Church to drop its opposition to the practice."[14]

"Drop its opposition to . . ." The reporter was dumbfounded. "With all respect, sir, the government has no right to meddle in the doctrines of Holy Mother Church!"

"It does now," the president replied. "Did you think our partnership would be a one-way street?"

The crowd, so recently united in support of a Christian nation, was now equally united in opposition to the idea.

* * *

This hypothetical tale uses Catholicism, the largest single denomination, as the religion of choice. But the story would run much the same regardless of the religion. Every denomination or perspective would isolate and marginalize a majority of the U.S. population. Catholics would find it every bit as intolerable to live under a state aligned with Jehovah's Witnesses, who would ban all Catholic saints and icons as idolatrous. Reform Jews would blanch at the social positions of a Southern Baptist regime—not to mention the belief that they themselves are bound for eternal damnation. Methodists under an Islamic state, Muslims under an atheist state—every perspective brings with it some very specific assertions, traditions, practices, and beliefs that run directly counter to a dozen others. Every perspective would result in a scenario similar to the one played out above.

Equally untenable is the myth that a "nondenominational" faith could be our public religious expression. There is no umbrella of faith that would not do fatal violence to the religious traditions of millions of Americans—not to mention more than 70 million who are unaffiliated with religion. The moment God is mentioned by name, Orthodox Jews (who never say or write "God" unless in prayer) are left in the cold, and the reference to a single God excludes millions of American Hindus and other polytheists.

Whenever someone advocates such an imagined umbrella, the details in reality—the name for God, the form of public prayers, and so on—bear a striking resemblance to their own perspective.

But isn't it a bit far-fetched to suggest, as the story above did, that religious organizations might actually *lead* the way for secularism? Not at all. Many denominations that now agitate for more religion in government and civic life were once the strongest advocates of strict separation of church and state.

## BAPTISTS FOR SEPARATION

When the Southern Baptist denomination was founded in 1845, it represented a very small minority, and it didn't want a majority vision of God forced on its adherents. So the Southern Baptist Convention included strong support for the separation of church and state and freedom of religion into the Baptist Faith and Message, the foundational statement of doctrine for the new denomination:

> Church and state should be separate. The state owes to every church protection and full freedom in the pursuit of its spiritual ends. In providing for such freedom no ecclesiastical group or denomination should be favored by the state more than others. Civil government being ordained of God, it is the duty of Christians to render loyal obedience thereto in all things not contrary to the revealed will of God. The church should not resort to the civil power to carry on its work. The gospel of Christ contemplates spiritual means alone for the pursuit of its ends. The state has no right to impose penalties for religious opinions of any kind. The state has no right to impose taxes for the support of any form of religion. A free church in a free state is the Christian ideal, and this implies the right of free and unhindered access to God on the part of all men, and the right to form and propagate opinions in the sphere of religion without interference by the civil power.[15]

The Southern Baptists fought hard to be sure public officials and public schools never endorsed any form of religion. Such decisions were matters of conscience, they said, to be expressed freely in the home and church.

Once Southern Baptists became the largest Protestant denomination in the United States, however, the value of separation was largely forgotten. Baptist churches and leaders became some of the most egregious violators of church-state separation, talking politics in the pulpits, demanding a return to Christian prayer in public schools, and lobbying for public policies that match their values.

Not all Baptists have forgotten. The Baptist Joint Committee for Religious Liberty, a nonprofit advocacy organization in Washington,

DC, is one of the most powerful and articulate defenders of separation in the country today.

## CATHOLICS FOR SEPARATION

One possible revision to the story: if the Catholic reporters were well versed in their own history, they might have rejected the mantle offered to them. American Catholics have been at the forefront of efforts to retain and preserve the strong separation of church and state—and for very good reasons.

Maryland, for example, was founded as a refuge for Catholics driven out of Europe by the religious wars of the seventeenth century, and out of other colonies by intolerant Puritan and Anglican majorities there. Instead of responding to these acts of intolerance with more intolerance, Maryland instituted the Maryland Toleration Act of 1649, ensuring religious liberty for all—well, at least all Trinitarian Christians. But Puritans seized control of the colony briefly in 1654 and by the 1680s represented a sizable majority of the population.[16]

Catholics remained a minority not only in Maryland but in the country as a whole. Despite the premise in the story above, any officially imposed version of God and religion is more likely to have sprung from somewhere in the Protestant coalition of faiths, and Catholics might well have found themselves once again in the position that had them fleeing Europe and the Anglican colonies in the seventeenth century.

Anti-Catholic attitudes continued well beyond the colonial period. Many states barred Catholics from holding office, and as Catholic immigration increased in the nineteenth century, lynch mobs killed Italian immigrants, arsonists burned down Catholic churches, and politicians railed against "the Catholic menace."

The anti-Catholic wave reached a zenith in the early twentieth century. A weekly newspaper called *The Menace*, devoted entirely to anti-Catholic agitation, reached a circulation of 1.5 million in 1915, larger than the largest papers in New York and Chicago. "The cowardice of a Roman thug has no parallel in either the human or animal kingdom," the paper frothed in 1914, calling for "men with red blood in their veins" to defend women and children from Catholics. "If we are compelled to live in this county with Romanists, as our weak-kneed Protestant critics say we are, the Romanists will have to be taught their place in society."[17]

So it should come as no surprise that U.S. Catholics have historically been at the forefront of efforts to strengthen and maintain the separation of church and state, since an official Protestant identity could only further marginalize their own faith.

## What Happens When the Wall Crumbles?

History is full of cautionary tales regarding the separation of church and state—and we don't have to go back 700 years to the Inquisition. The founders of the United States had only recently departed a country and continent convulsed by conflicts over which religious view would wear the crown of state—conflicts that damaged not only the state but the churches as well, not to mention countless individuals.

Governments allied with a single religious perspective have a long history of turning a blind eye to popular uprisings against other religions. When Anglican mobs burned the homes of religious Dissenters in the "Priestley Riots" of 1791 in Birmingham, England, the Anglican government of William Pitt was slow to respond. This appears to have been intentional, as local officials were later found to have been involved in planning the attacks.[18] Similar "blind eye" responses by religiously aligned governments were seen in anti-Jewish pogroms in the nineteenth and twentieth centuries,[19] and more recently in the government of Myanmar's response to attacks on the Muslim minority by Buddhist extremists.

This is not to say that countries without an official or dominant national religion are completely devoid of interreligious conflict. The U.S. anti-Catholic violence mentioned above, and state laws banning Catholics, atheists, and Jews from holding certain offices— many of which are still on the books—show that such conflict does indeed occur. But when the state is neutral in religious identity, there is a greater chance of quelling such outbursts of violence and discrimination because the government can act as an impartial arbiter, enforcing the law and pointing to its own legal mandates for tolerance and plurality. This also means that no religion needs to be fearful of a program of state persecution. This allows religion to soften—to become less fanatical and more tolerant.

## CULTURAL EVOLUTION: THE ENGINE OF HUMAN PROGRESS

Ever since human genetic evolution achieved a leap in cognitive function roughly 70,000 years ago,[20] *Homo sapiens* has been able to revise its behavior rapidly in accordance with changing needs. This opened a fast lane of cultural evolution, bypassing the traffic jams of genetic evolution.[21]

Evolution at the genetic level moves at an unimaginably slow pace. It can take thousands of years or more for a single beneficial mutation to start showing up at the population level. Fortunately, another kind of evolution can benefit humanity much more quickly. Through the sharing of art, stories, education, and advancement in science, humans are making progress through cultural evolution that promotes knowledge and spreads values that improve human life. That's why life is better, longer, and safer now than ever before.

That statement may come as a surprise. The press and social media are quick to convey news of violence and disaster, and many of us are convinced that life is a great deal *less* safe and more violent now than in the past. But it's not true. The irony is that despite this ever-heightened state of fear, most people in both developed countries and undeveloped countries enjoy a level of physical safety unprecedented in human history.[22] Despite the constant claim that violent crime has never been worse and the perpetual warning that "you can't be too careful in this day and age," the U.S. Bureau of Justice Statistics reports that as of 2013, violent crime in every category has dropped sharply for twenty years and reached the lowest point since records have been kept.[23] (See figure 1.) Our collective behavior has improved over the centuries as a result of cultural evolution on many fronts that both protects us from harm and improves our emotional and intellectual well-being, including:

FIGURE 1. CRIME RATES VS. PERCEPTION

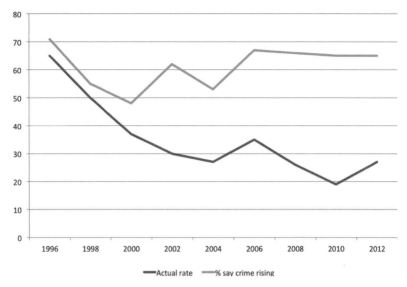

*NOTE*: VIOLENT CRIME RATE IS VICTIMIZATIONS PER 1,000 HOUSEHOLDS
*SOURCE*: BUREAU OF JUSTICE STATISTICS, NATIONAL CRIME VICTIMIZATION SURVEY, 1996–2012.

- reduction in human and animal sacrifice;
- reduction in warfare and violence;
- a marked reduction in slavery;
- less subjugation of women;
- protections for children;
- development of science, medicine, and agriculture;
- development of law, courts, and justice;
- creation of international diplomatic bodies (such as the UN) and tribunals (such as in The Hague);
- development of language, writing, literature, storytelling, and gossip;
- technology for communication at a distance, including telephone, radio, television, the Internet, and audio and video recording;

- development of cooperative participatory arts, including music and dance.

How can we assist this process, speed it up, and spread it to all parts of the world, including the entire Middle East and Africa?

A crucial step in the progress of cultural evolution is greatly enlarging the circle of our empathy. We live in naturally concentric circles of community, representing what has been called an "empathy gradient." The inner circle is often our immediate family, those with whom we feel the most connection. Enclosing that circle is the larger circle of extended family, followed by other affinities like regional, cultural, and language groups, our ethnicities, regions, nations, religions, our worldviews, even our species. Each of these circles defines a community, linking us to those with whom we share something significant.

Genetic evolution has given us a strong sense of empathy and moral care toward people within our own families and tribes but little sense of empathy toward people of other groups. As for those with whom we feel the least in common—the concentric circles at the furthest remove—genetic evolution has inclined us toward fear, distrust, and outright conflict.[24] (See figure 2.) A major challenge for cultural evolution is to overcome or offset the naturally evolved genetic tendency toward tribalism that continues to cause considerable suffering and death, based in part on religious tribalism.

An important contributor to the problem of tribalism derives from the "silo effect"—the tendency of different groups to cut themselves off from outside input, preferring to hear and repeat the same assumptions, ideas, and perspectives as the like-minded group members around them. Such groups can develop a shared understanding of the important "facts" of reality that differ from the actual facts.[25] Most people are either born into a given group— religiously, culturally, or otherwise defined—and accept the "facts" of that group, or they later choose to join a group and accept its "facts." Differing factual beliefs can cause or exacerbate conflict between groups and between individuals such as within a family.

That is not to say tribal or group identities are always negative.

FIGURE 2. CONCENTRIC CIRCLES OF EMPATHY

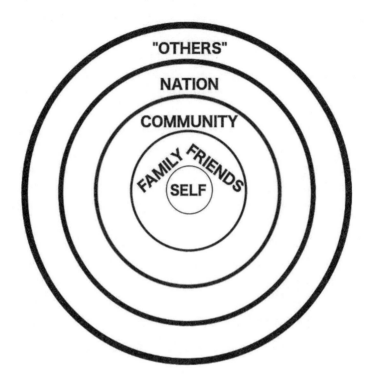

SOURCE: OWN WORK, CREATIVE COMMONS CC 0, PUBLIC DOMAIN

Sometimes a group can coalesce around shared positive values and actions. A church member who feels "called" by her denominational identity to live her life in the service of others and to join with others of her group in alleviating poverty is responding to a positive tribal incentive. The Buddhist who feels similarly inspired by the example of the Buddha or a *bodhisattva* to relieve the suffering of others, or a humanist inspired by empathy and compassion for others to feed the hungry and house the homeless—in each case, the individual is responding positively to the values of his or her tribal identity.

There are benefits for both the members and for society when people come together in congregations to discuss and listen to each other about ways to think and behave, whether those congregations are religious or other. Because people are social animals, when they share ideas on how to think and behave, extremes and antisocial attitudes are moderated back toward the center, which tends to reduce strife.[26] And there is much to think about and discuss. No one always behaves as their own ideals suggest they should, leaving people with a sense of guilt – should I spend money on myself or should I use it to help the poor? Religion-like congregations can help people manage their sense of guilt rather like priests claiming to forgive sins.[27]

But when different tribalistic identities lead to the acceptance of fundamentally different factual assumptions about the world, the result is far more troubling, and sometimes far more dangerous.

If all people can agree on the important facts of reality, human culture will be primed to move forward at all levels, from family and friends to groups and nations. The only way all people can agree on facts is by accepting the conclusions of science—the only reliable fact-determining engine in our solar system.

## WHY PROMOTE ACCEPTANCE OF SCIENCE?

The word "science" is derived from *scientia,* Latin for knowledge in a general sense. In older usage, what we now call "science" was more accurately called *empirical science*—knowledge that has been tested by rigorous experiment. Even when so tested, scientific knowledge is always held tentatively and subject to disconfirmation.

Most of what people hold as knowledge is not arrived at by this rigorous process, and rarely held as tentatively as it should be. The result of this inflexibility of beliefs can be dangerous. If people believe that a certain set of "facts" is beyond question, and that those "facts" make it acceptable (or even obligatory) to oppress or kill people of other religions or sects or tribes,[28] they are not easily persuaded that their "facts" are wrong. We need to help them recognize the benefit of reversing their beliefs in those false "facts." Many Islamic extremists, for example, believe their god gives them the right to kill people who do not accept their form of religion or who disrespect their religion. Many Christian conservatives in the United States believe that the "truth" as revealed by their god gives them the right to limit access by anyone to abortions, even by people who do not share their religious convictions.

So does this mean science and religion are utterly separate spheres, irreconcilable in any way? It often feels this way today. But before the Scientific Revolution of the seventeenth century, religion shaped and guided much of our inquiry into the nature of the world around us. That the nature of religion itself often compromised that inquiry and produced less than reliable results is beside the point. The same can be said about early scientific methods. It was still inquiry, still an attempt to interrogate reality, which is why theology was formerly known as the "Queen of the Sciences." The problem is not the goal, but the tendency toward dogmatic thinking that leads to absolutes, closed to the possibility of disconfirmation.[29]

To dislodge false beliefs, we need to promote acceptance of the only effective fact-determining engine yet devised—science. Science is simply the collective term for *all* methods that have been found over time to be reliable in determining facts about the world, in addition to the facts that have been assembled.[30] Many people find it hard to learn and understand science. However, what people need to understand about science can be distilled to make it easy for anyone to learn the essential parts, which we explain below.

We can't effectively spread secularism to everyone without spreading acceptance of the scientific way of knowing and the scientific consensus on the facts. People who have a false understanding of important facts will tend to generate disputed values from those false "facts" and will often try to force their disputed values on others,[31] which violates a principle of secularism. They will deny that they are violating a core value of secularism because they claim to be acting on the truth. To put it simply, everyone is entitled to their own opinions on values, but not their own facts.

The benefits of promoting secularism and science extend beyond the problems of religiously motivated oppression. They also address a host of other problems, including death and disease from willfully ignoring or denying health science and environmental destruction.

A sobering illustration of such a problem occurred in December

2014 when a person infected with measles spread this disease to at least 78 people at Disneyland in California.[32] Of the 34 measles victims who resided in California and whose vaccination history could be ascertained as of January 23, 2015, 28 (82 percent) had not been vaccinated for measles. The U.S. Centers for Disease Control and Prevention reported that, of all measles cases in the United States in the first five months of 2014, 74 percent occurred in people who declined vaccination "because of religious, philosophical or personal objections."[33]

This is a stark example of the harm caused by verifiably false beliefs. The science is clear: the measles vaccine is both cost-effective and health-effective. The vaccine confers lifelong immunity in 99 percent of people who receive two doses. An outbreak in the United States in 1989–91 caused 55,000 infections, 10,000 hospitalizations, and 123 deaths.[34] In response, measles vaccination then rose to a 95 percent rate, and by the year 2000 measles was eliminated from the U.S. population.[35] However, an antivaccine movement began with the unfortunate publication of inadequately peer-reviewed scientific papers in 1998 and 2001 that linked vaccines to a rise in cases of autism. Although the claims of these papers were retracted, journalists and celebrities seized upon the controversy. Fear is an effective way to get media coverage, but it can also trigger a panic. In some communities, so many parents refused to vaccinate their children that they gave measles a path to spread, which allowed the Disneyland outbreak and other outbreaks to happen.

Popular rejection of the science of vaccination causes avoidable spreading of disease (see figure 3), including to the 1 percent who do not develop a permanent immunity from the vaccine and the 1 percent who cannot physically tolerate the vaccine. People who do not vaccinate their children are avoiding their fair share of vaccination risks and taking a free ride on the responsible citizenship of those who do. The problem is global. In 2014, measles caused 115,000 worldwide deaths that could have been prevented.[36] We can save lives by spreading acceptance of the scientific consensus on vaccinations.

Rejection of the measles vaccine is just one example of the importance of accepting the scientific consensus to determine all

FIGURE 3. NUMBER OF U.S. MEASLES CASES, 2001–2016

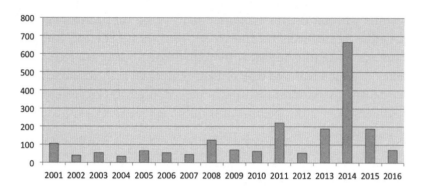

*SOURCE*: CENTERS FOR DISEASE CONTROL AND PREVENTION, HTTPS://WWW.
CDC.GOV/MEASLES/CASES-OUTBREAKS.HTML

facts. Consider that in the United States from 1975 to 1995, at least 169 children died from diseases with a 50–90 percent cure rate because parents denied medical treatment for religious reasons.[37] In a single year in the United States, 2012, there were at least two cases where such children died of appendicitis, which has a greater than 99 percent medical treatment success rate.[38]

Our goal here is to articulate a campaign to educate the world not about science broadly, as that would be too ambitious, but about the power of people simply accepting the scientific consensus on the facts and understanding the imperfect nature of knowledge. We seek to get different tribes to accept the same facts,[39] and thus reduce the mutual "otherness" that leads to conflict. We seek to promote individual happiness through knowledge.[40] We need to counter the science deniers and spread acceptance of the conclusions of science on all topics that impact human well-being. Spreading acceptance of science offers greater benefits for mankind than spreading of atheism. If people want to hold on to a belief that there is or was some kind of god or source of morality outside of humanity that did or does something, this lingering unscientific belief will cause relatively little harm compared to other science denials.

We hope to make the case not just by "preaching to the choir," i.e., to those who accept secular separation of religion and politics, but by making the case to all members of our society and the world. And we need to do so effectively, which (to paraphrase Neil deGrasse Tyson) includes a sensitivity to the state of mind of those who oppose vaccinations or fluoridation of water supplies and other developments of science that reduce suffering. It is the facts plus the sensitivity that, when convolved together, create impact.

## NATURAL AND INTUITIVE WAYS OF KNOWING

Not all knowledge results from education. A crucial feature of our own evolution has been the development of intuitive ways of knowing. These range from the useful disgust response that keeps us from eating rotten meat to pattern recognition that allows each person to quickly assess their environment and make decisions. The beliefs of early humans would have formed individualized worldviews, but eventually language gave us the ability to share our beliefs, leading to shared worldviews with lots of false beliefs.

Sometimes the shared beliefs are effective and true, and language gives us the ability to share acquired wisdom. But some conclusions are accepted and spread not because they are effective and true, but because they are emotionally preferred. Reaching emotionally preferred answers quickly becomes a hard habit to break. Natural inclination, reinforced by indoctrination, can and does lead people to misunderstand the causes of illness, predation, and other risks to life and limb. Having misunderstood the cause, we are easily misled about the best response to a problem. And when the disconnect between belief and reality fails to cure the sick child or quiet the storm, instead of adjusting our beliefs, humans have an unfortunate record of scapegoating one another to appease the forces we imagine are in control of our destinies.

Our very helpful and adaptive tendency to find patterns in the world around us sometimes misfires, and we give meaning to patterns that are in fact meaningless. The rain that falls on a drought-ridden village after a "witch" is burned may seem to the villagers like convincing evidence of the efficacy of witch-burning for drought relief. Never mind that a half-dozen previous burnings had no effect. The belief provides a powerful illusion of control over otherwise terrifying natural forces. Our incentive to cling to the belief is therefore extremely high.

Over time, such beliefs, reinforced by false pattern recognition, can result in entire false systems of thought such as astrology, energy healing, and karma. With enough time and emotional incentive, such beliefs can gain a foothold in a specific cultural space.

It is easy to say we simply shouldn't believe things that aren't true. But the incentives are not to be taken lightly. As the poet and author Jennifer Michael Hecht once said, the central difficulty of our existence is that "we are human and the universe is not."[41] Lacking a heart to which we can appeal, the universe can seem an unbearably cruel home, especially to those who are relatively powerless. Actual knowledge has proven our most effective remedy for this powerlessness, the best way for us to bring a capricious world under our control. But the emotional pull of our intuition-fueled beliefs is tremendously strong. And this is no surprise, since evolution tailored them to meet our every need.

Only by empowering people, by lifting them out of ignorance and fear, can the strength of real knowledge begin to eclipse the power of traditional or intuitive false beliefs. The most effective way to achieve that empowerment doesn't start with ridicule, which often causes people to retreat further into the ridiculed belief. It begins with empathy for those who still find themselves in the grip of those natural beliefs.

## THINKING POORLY, THINKING WELL

Critical thinking can be defined as the systematic attempt to avoid errors in reasoning. We want the right answer, of course, but our natural tendencies often incline us toward false "facts" that satisfy our preferences. Learning to avoid those errors is an important step toward finding the truth. Fortunately, the errors we make are so predictable that some are instantly identifiable. As a group, they are called logical fallacies, and getting to know them can help us avoid them.

Here are a few of the more common fallacies:

**Post hoc ergo propter hoc:** Because one event followed an earlier event, the earlier event caused the later one (e.g., a baseball player won a game after drinking a particular beer, so he should drink the same beer before every game).

**Affirming the consequent:** Reversing the terms of a conditional statement (e.g., "All crows are black, this bird is black, therefore this bird is a crow").

**Straw man:** Refuting a simplified or otherwise inaccurate version of an opponent's position rather than the position itself.

**Cherry picking:** The careful selection of only those data that support one's preferred conclusion while ignoring those that refute it.

More than 200 of these common thinking errors have been identified and defined. But two tendencies underlie nearly every error. They are called confirmation bias,[42] the natural tendency to see events in ways we prefer, and motivated reasoning,[43] which drives us to arrive at conclusions that match our preferences. More than anything else, the methods of science and critical thinking have been created to control for these two errors.

Confirmation bias is not only a feature of "bad thinkers"; it is an intrinsic part of our factory settings as human beings. You and I and everyone we know have a strong tendency to see what we want to see. It is amazing how reliably I notice that my kids are the most remarkable performers in the talent show or the most impressive athletes on the playing field. Of course, I tend to notice the things that confirm my opinion (the completed goal or successful high note) and forget the ones that contradict it (the bad pass or sour note). That is why I'm a terrible choice to judge their talent shows or referee their games: confirmation bias impairs my judgment, tilting me in the direction of the conclusion I'd prefer—that my kids are recognized as the best.

Though scientific methods are intended primarily to control for confirmation bias and motivated reasoning, individual scientists are no less prone to these errors. If they want to be taken seriously in the overall enterprise of science, they are required to submit to a system that controls for their biases, including procedures such as the double-blind study where appropriate and peer review, designed to take the straw of scientists' human work and spin it into the gold of science.

Conspiracy theories, the very embodiment of confirmation bias and motivated reasoning, work in the opposite direction. They use what Thomas Eagar of M.I.T. called "the reverse scientific method: Determine what happened, throw out all the data that doesn't fit your premise, then hail your findings as the only possible conclusion."[44] (This illustrates another logical fallacy called the Texas Sharpshooter, which imagines someone shooting at the side of a barn, then painting a target wherever the most hits landed.) The key to avoiding confirmation bias and motivated reasoning is to put yourself at the critical mercy of people who don't share your biases. It can be a painful experience, but it is both the surest way to gain credibility and the surest way to discover the real world.

## FACTS ABOUT FACTS

### THE BEST WAY OF KNOWING FACTS IS THE SCIENTIFIC WAY.

As we have seen, the evolution of human culture has produced a superior way of knowing facts called "science." This way of knowing does not come naturally for humans. It requires education—learning from the pool of knowledge assembled by the collective efforts of others over hundreds of years. It is a collaborative process. No "fact" is added to the collection of scientific knowledge until it has been vetted and verified by many people engaged in the scientific process.[45]

Knowing the truth isn't always important. When you consider eating food and it smells repulsive, your intuition shouts that the food is unsafe to eat. You rely on your intuitive way of knowing and do not check it against scientific knowledge. It could actually be food that is safe and considered a delicacy in another culture.

But there are many situations where intuitive knowledge must give way to scientific knowledge to minimize human suffering. When a deadly outbreak of cholera decimated the Soho district of London in 1854, a full understanding of germ theory—the idea that diseases are often caused by microorganisms—was still roughly a decade away. In just three days, 127 people died of cholera within a few blocks in the heart of the city; within ten days, the number was 500. Then, a medical inspector named John Snow used scientific methods to isolate the cause of the outbreak.

By interviewing the families of victims, he discovered that almost every one of the dead had consumed water from a single well at the juncture of Cambridge and Broad Streets. A map of the victims' locations provided further evidence: the 500 victims were almost all within a 250-yard radius centered precisely on the Broad Street pump (see figure 4). Only five of the deaths occurred in closer proximity to another pump. Snow's solution was simple: he removed the pump handle. Within days, the epidemic was over.

Through subsequent interviews, Snow learned that the mother of a baby suffering from cholera had washed the child's diaper into a basement cesspit just three feet from the pump. The cesspit leaked into the well water, transmitting the disease to those who consumed the water.

In this case, knowing the truth was literally a matter of life and death. A scientific approach had isolated the cause and suggested a solution. But confirmation bias reared its head in the response of city officials who publicly rejected Snow's theory and replaced the pump handle. The idea that fecal matter was involved in the outbreak was considered too foreign, and disgusting, to accept.

The scientific way of knowing facts does not dismiss contributions from the "humanities" as contrasted with the "sciences," provided those contributions are not based on false factual beliefs. There is no clear boundary between scientific and humanistic scholarship. What we call the "sciences" and what we call the "humanities" is largely a matter of practicality.[46] Any method, including humanistic methods, that works to achieve a broad, peer-reviewed consensus on objective facts is valid as science. The humanities provide the most effective way to address the subjective experience of the human condition—the

FIGURE 4. MAP OF CHOLERA OUTBREAK IN LONDON, 1854

essence of being human—while the sciences are limited to objective facts.[47]

Of course, everyone must make decisions every day based on intuition, inspiration, or tradition with no scientific basis. When you sit down, you rely on your intuition that the chair is real and can hold your weight. The scientific way of knowing does not suggest that the natural and intuitive ways of knowing facts are useless; they are necessary to all life functions. But people who accept the scientific way of knowing remain ready, upon learning the scientific consensus on a topic, to amend their beliefs based on intuition or tradition and replace them with current scientific thinking.

Few people if any practice the scientific way of knowing perfectly. We are all affected by motivated reasoning and confirmation bias. The best we can do is honestly try to employ the scientific way of knowing the facts in our decision-making when it matters most, and to lay ourselves open to cross-checking and critique by those who do not share our immediate biases.

### SPREADING ACCEPTANCE OF THE SCIENTIFIC CONSENSUS ON THE FACTS WILL ADDRESS RELIGIOUS EXTREMISM MORE EFFECTIVELY THAN SPREADING ANY OTHER CULTURAL IDEA.

The world's major religions all assert false "facts" and then base their values at least in part on these false "facts." Though these "facts" are persuasive to those within a given belief system, they carry little weight for adherents of other belief systems. To people outside of a given belief system, the asserted facts often seem ridiculous.

Consider the views of mainstream Christians on the "facts" asserted by the Church of Jesus of Latter-Day Saints—that Mormon founder Joseph Smith was visited by an angel in upstate New York in 1820, that the angel directed him to buried golden plates describing (among other things) a supposed visit by Jesus Christ to North America, that the dead can be baptized into the Mormon church by proxy and informed of this event by spirit messenger, that three heavens exist, and more. Many Christians consider these beliefs not only false but bizarre. Yet non-Christians tend to see the beliefs of mainstream Christians—that a god was incarnated in a human body, died, and was resurrected, that Christ is physically present (as Roman Catholics believe) in communion bread and wine, that nonbelievers are consigned to eternal damnation after death, and so on—as equally strange and untenable.

It is difficult for people of traditional religions who personally hold moderate and tolerant views to persuade religious extremists that their facts are wrong, because the tolerant or moderate factions accept the same source or a similar source as containing revealed genuine facts. The moderates merely give the words a more modern interpretation, which, as often as not, is inconsistent with the plain meaning of the words.[48] This brings the conversation to something of an impasse, with each side claiming the other has misinterpreted the meaning of the source, and neither is able to proceed on any objective basis. What is needed is an objective system external to the religion itself to which moderates can appeal.

Religious moderates and the nonreligious generally agree that religious extremism is causing very real suffering in the world. Sadly, the chances are small that people of goodwill but with a commitment to traditional religions that teach one set of dubious facts will ever persuade extremists of their own religion or any other religion to stop believing another set of dubious facts. Arguments based on science are the only arguments with any potential to persuade religious extremists that their "facts" are wrong.[49] If we are to save humanity from naturally evolved tribalism expressed as religious extremism, people who fully accept the conclusions of science, including religious moderates, must try to persuade the religious extremists to accept their secular and scientific views.

Any other approach boils down to this: one group that relies for some of its facts on a traditional interpretation of religious beliefs or on texts recorded in pre-modern times claims that another group is wrong to rely on a different interpretation of the same or another ancient text. In effect, each group says to the other: "Your delusions are wrong; my delusions are right!"

## To Build Good Values, Start with Genuine Facts.

Determining what values each person should live by is the most important project of human cultural evolution. These values have been evolving in each society since the beginning of human culture.[50] How to determine what values are best is a difficult topic beyond the scope of this book. We decline to comment on what values people should adopt other than the value of discovering and accepting truth and the three core values that are essential to secularism:

1. Equality of all religious/tribal/ethnic groups in all aspects of government.

2. Liberty / Freedom of individual conscience except where required to avoid harm to others.

3. Truth as the basis of all government decisions.

The value of discovering and accepting truth in all matters is an extension of the third value of secularism, extending it from merely government decisions to all aspects of life and perception of the world. For reasons explained below, we call this value *evidism*.[51] We are arguing for acceptance of only four values: the three core values of secularism, which are required to minimize human conflict, plus *evidism,* which is required to move human cultural evolution forward to new heights. As we show below, it is an objective fact that universal acceptance of the core values of secularism is the only achievable way to have peace between humans, and universal acceptance of *evidism* is essential to fully implementing the core values of secularism.

To examine the relationship between facts and values, we can look to the long history of racial inequality. For centuries, the "fact" that an unbridgeable genetic gap in ability and intellect existed between races was widely accepted. As evidence accumulated to the contrary, in the form of scientific research and individual experience, the confirmation of a lack of relevant genetic difference led to the ethical belief that all individuals should be treated the same way. This is not to suggest that we all jumped into the boat once the facts were in, or that we have all jumped into the boat even now. It has been a long, difficult struggle, complicated by bias and ignorance, and it is by no means over. But this is just one example of values being increasingly constrained by discovered empirical facts.

The idea that there may be superior values that can be derived from facts, which values should be spread to all cultures, is attacked from both sides of the political spectrum for different reasons. From the right, the argument against facts determining values is primarily religious. It centers on the perceived need for grounding such values in something beyond human reason. We are too deeply flawed, goes the argument, to be trusted with anything as important as goodness. From biblical accounts like the story of Abraham and Isaac to recent parables like *Lord of the Flies*,[52] such ideas are rooted in a deep distrust of the human capacity to generate our own moral understanding. On the left, the main discomfort is with the suggestion that there might be a reliable concept of right and wrong that transcends all cultures, as science itself does, which implies that some cultures have "better" values than others.

Some people argue that whether people have a true understanding of facts is not important; we should merely try to lead them to good values and let them believe any "facts" they like. It is certainly true that the values people act upon are far more important than the facts that people believe. For example, a person can accept all the conclusions of science in the solitude of his mind and still hold a value of denying equal rights to people of other tribes or religions or races. Determining what values are preferable is not easy under the best of circumstances, and it is difficult to persuade people that particular values are better than others. Except for the values of secularism and evidism, there are no objective standards for finally deciding which values are best.

By contrast, there *are* objective standards for determining facts. These objective standards are what we call *science*. Large numbers of people claim to base their values on beliefs about facts that are demonstrably wrong. If people accept the conclusions of science, it is easy to reach agreement on all facts of consequence, and easier to set aside poorly founded values based on false "facts." Getting the facts right will reduce the challenge of leading people to agree on values.

But using science to achieve universal agreement on facts will not automatically lead people to agree on values. Some have argued that, when people abandon traditional religion due to conflicts with science, they naturally adopt the values of humanism, a philosophy that commonly replaces religion and includes no supernatural beliefs. While this may be true of a majority, it does not automatically follow. All statements of what it means to be a humanist that have been published include particular values.[53] People who agree on all the important facts by adhering to science can still disagree on many values, including politics, economic systems, punishment of criminals, education policy, polygamy, prostitution, gender equality, abortion, contraception, environmental issues, treatment of animals, proper eating practices, and more. Getting the facts right is necessary for selecting good values, but it isn't sufficient.

Still, that crucial step advances the project of human happiness profoundly. If we can agree on the important facts of reality by accepting the conclusions of science (*evidism*), it will likely increase human happiness in a number of ways:

1. Adoption of a shared understanding of the facts by an increasing number of people would likely reduce future conflict. Accepting the conclusions of science is the only way humans can ever reach agreement on the facts, because many people will never accept any asserted facts that are known by science to be wrong.

2. Individuals are invariably better off navigating through life when they achieve a correct understanding of the dividing lines between reality and fantasy and are no longer confused about whether there is factual truth or whether people can ever know the truth. A tendency to doubt the accessibility of knowledge often results from being taught falsehoods.

3. Increasing education in humanities increases people's abilities to understand one another's viewpoints and values and reduces conflict.[54] Spreading acceptance of the conclusions of the humane sciences about people also increases our abilities to understand one another's viewpoints and values.

Each person moving to accept the conclusions of science will make a major contribution to the positive evolution of human culture and reduction of suffering. This will require evolution of each religion to accept the scientific consensus on all important facts, even as the religion retains other aspects of its values, culture, and tradition.

You may feel yourself pulling away at this point, shaking your head, convinced we are proposing a pie-in-the-sky idea that simply can't be done. And who could blame you? The history of the intersection of science and religion is hardly encouraging. But that history to date has largely been written from two extreme perspectives: the false assumption that there is no conflict between religion and science and the false assumption that religion is fixed and cannot change without being destroyed. This book explores a third option: an evolution of our discourse, and of religion itself, to bring the two spheres of human experience into a cooperative relationship without destroying either religion or science.

Just as several religions have evolved to accept the principles of secularism, religions can also evolve to accept the scientific consensus on all facts (evidism). We know this is possible because, as we detail below, several religions have evolved to do exactly that. The typical path is first to accept all of science except for the conclusion that there is no sufficient evidence for any belief in a god. Achieving this step provides huge benefits. The final step is for them or their children to finally dispose of this last unscientific belief.

This book presents ideas for adjusting our language and messages to make them most effective for spreading the core values of evidism and secularism to our friends, family, neighbors, countrymen, and fellow citizens of the world. It also presents specific teachings that we can offer to advance the principles and practice of evidism and secularism.

Successfully spreading secularism and evidism requires in many cases an understanding of how we can effectively communicate with one another. What Neil deGrasse Tyson calls a sensitivity to the state of mind of your audience is a vital part of effectively sharing reality.

## NOTES

1. Russell Blackford, *Freedom of Religion and the Secular State* (Blackwell Public Philosophy Series) (Malden, MA: Wiley-Blackwell, 2012), 11, 12.

2. In the United States and Europe, in recent decades, the percentage of people not affiliated with any religious group has been growing dramatically, and new social organizations have been growing. Some of these new social organizations call themselves "religions," as discussed below, and some do

not. However, the Western world is still far from devoid of religion and other parts of the world are far behind these trends. To date, there is no credible basis to predict that "religion" will ever fade from the world. We are predicting that many religions will change to accept all science, and some of these will be with us long into the future.

3.    Steven Pinker, *The Better Angels of Our Nature* (New York: Viking, 2011).

4.    In his 1996 book, *The Clash of Civilizations,* Samuel Huntington describes these religious conflicts as "clashes of civilizations."

5.    For a thorough argument that using "secular" in this newer and narrower way is hurting the cause of secularism, see Jacques Berlinerblau, *How to Be Secular: A Call to Arms for Religious Freedom* (Boston: Houghton Mifflin Harcourt, 2012; Kindle Edition).

6.    On this topic, see *Secularism and Freedom of Conscience* by Jocelyn Maclure and Charles Taylor (2011) and *Liberty of Conscience* by Martha Nussbaum (2008). Maclure and Taylor refer to the first two core values as "foundational principles" of secularism. Granting freedom of conscience to minority groups with differing sets of values, e.g., freedom for Native Americans to use mind-altering drugs, is called "values pluralism." See William Galston, *Liberal Pluralism: The Implications of Value Pluralism for Political Theory and Practice* (2002).

7.    http://www.secularism.org.uk/what-is-secularism.html. Ronald A. Lindsay defines secularism as "the view that: [1] government should not involve itself with religious matters, [2] religious doctrine should play no role in shaping public policy or in the discourse about public policy, and [3] religious institutions and beliefs should not enjoy a privileged position within society." *The Necessity of Secularism* (Durham, NC: Pitchstone Publishing, 2014), 18. There are authors who use the word "secularists" to refer to people who reject beliefs in anything supernatural or transcendent— i.e., the opposite of "religious." See, e.g., Philip Kitcher, *Life After Faith: The Case for Secular Humanism* (New Haven, CT: Yale University Press, 2014; Kindle Location 365). We argue below that this is a counterproductive use of the word *secularist.* We should use the word *secular* in a way that leaves room for all people who are religious to be *secular* as well. For a thorough argument in favor of our view, see Berlinerblau, *How to Be Secular.*

8.    As Ronald Lindsay explains in *The Necessity of Secularism,* 183, there is no need to persuade others out of their religion; we only need to convince them of the wisdom of secularism.

9.    Lindsay, *The Necessity of Secularism.*

10. One of many strong statements of Catholic support for church-state separation is contained in *Dignitatis Humanae,* a declaration by the Second Vatican Council in December 1965. Many Islamic scholars claim that secularism is the best way to follow Islam and the best way to implement *sharia.* http://en.wikipedia.org/wiki/Islam_and_secularism.

11. "XVII. Religious Liberty. God alone is Lord of the conscience, and He has left it free from the doctrines and commandments of men which are contrary to His Word or not contained in it. Church and state should be separate. The state owes to every church protection and full freedom in the pursuit of its spiritual ends. In providing for such freedom no ecclesiastical group or denomination should be favored by the state more than others." Full text available at http://www.sbc.net/bfm2000/bfm2000.asp.

12. For an excellent presentation on the importance of secularism, see Lindsay, *The Necessity of Secularism.*

13. The Openly Secular project has published works of many authors using this narrower definition of who is secular and who is not. http://www.openlysecular.org/. See http://admin.openlysecular.org/wp-content/uploads/2015/03/OpenlySecular_RiseOfTheNonreligious_March2015.pdf.

14. The United States Conference of Catholic Bishops announced opposition to the death penalty in 1985 and affirmed this position in 2005. http://www.usccb.org/_cs_upload/7917_1.pdf.

15. "Religious Liberty," Baptist Faith and Message Part XVII, available at http://www.sbc.net/bfm2000/bfm2000.asp. Accessed December 18, 2015.

16. Robert J. Brugger, *Maryland: A Middle Temperament* (Baltimore: Johns Hopkins University Press, 1988).

17. Ed Brayton, "When the Muslims Were the Catholics." Available at http://www.patheos.com/blogs/dispatches/2015/12/11/when-the-muslims-were-the-catholics/. Accessed December 22, 2015.

18. F. W. Gibbs, *Joseph Priestley: Adventurer in Science and Champion of Truth.* (Nashville, TN: Thomas Nelson, 1965).

19. Harvey Goldberg, "The Anti-Jewish Riots of 1945: A Cultural Analysis," *Jewish Life in Muslim Libya: Rivals and Relatives* (Chicago: University of Chicago Press, 1990).

20. Sometimes called the Cognitive Revolution or the Great Leap Forward, when human behavior changed rapidly in complexity and creative/practical focus, including advanced toolmaking, language development, and migration. See also Jared Diamond, *The Third Chimpanzee* (New York: Harper Perennial, 1992), 47–57.

21.  Yuval Noah Harari, *Sapiens: A Brief History of Humankind* (2014; New York: Harper, 2015).

22.  Steven Pinker, *The Better Angels of Our Nature.*

23.  http://www.bjs.gov/index.cfm?ty=pbdetail&iid=5113.

24.  Joshua Greene, *Moral Tribes* (New York: Penguin Press, 2013).

25.  Edward O. Wilson, *The Meaning of Human Existence* (New York: Liveright, 2014).

26.  In *Breaking Faith*, Peter Beinart reviews the decline in church attendance by both liberals and conservatives and suggests that this might be causing politics in the US to become more extreme on both the left and the right and might be a causative factor in the election of Donald Trump. Atlantic Magazine April 2017, https://www.theatlantic.com/magazine/archive/2017/04/breaking-faith/517785/.

27.  In an article in *The Hedgehog Review* entitled The Strange Persistence of Guilt, Wilfred McClay explores the sense of guilt that people have without any traditional religious origins. Human innate morality that we are born with because we are social animals leaves nonsociopaths always feeling somewhat guilty about their actions in society or failures to act. McClay posits that, without traditional religion to relieve this guilt with granted forgiveness, our society will face greater challenges for happiness and morality than when nearly everyone believed in a god that could forgive sins. http://iasc-culture.org/THR/THR_article_2017_Spring_McClay.php

28.  The Crusaders believed in false "facts" that made it right for them to kill or oppress Muslims. Islam takes this position today more than any other major religion (Samuel Huntington, *The Clash of Civilizations*). This is a major motivation for violence between Muslim groups such as Shias and Sunnis.

29.  The authors are indebted to theologian and Jesus Seminar member Lloyd Geering for sharing the text of "How Is Science Related to the Judeo-Christian Tradition?," his 2004 lecture to the Royal Society of New Zealand, from which ideas in this section are drawn.

30.  In *Every Thing Must Go: Metaphysics Naturalized* (Oxford: Oxford University Press, 2007; Kindle location 512), James Ladyman et al. state: "there is no such thing as 'scientific method.'" "Science is our set of institutional error filters for the job of discovering the objective character of the world." "Science is . . . demarcated from non-science solely by institutional norms," citing "peer review" and "representational rigor" as elements required by institutional norms and citing as examples of institutional prohibitions, do not invent data and do not take samples unrepresentatively. "With respect

to anything that is a putative fact about the world, scientific institutional processes are absolutely and exclusively authoritative." Some authors define "science" more narrowly and separate it from, for example, history, even when undertaken with full rigor. We find it more useful for marketing to let "science" stand for all rigorously assembled facts, recognizing that there is a unity of knowledge and no bright line between the narrow concept of science and any other rigorous inquiry. Some authors prefer to use "rational inquiry" as we are using "science."

31. A religion can be fundamentalist and full of science denial and still fully accept secularism. However, as weighted by numbers of people in each religion, this is rare. For example, the Catholic Church all over the world attempts to enact its view of morality into law, for example on the topic of abortion.

32. As of January 23, 2015.

33. http://www.cdc.gov/mmwr/preview/mmwrhtml/mm6322a4.htm?s_cid=mm6322a4_w; http://www.latimes.com/business/hiltzik/la-fi-mh-measles-20140530-column.html.

34. http://www.slate.com/articles/health_and_science/medical_examiner/2015/01/measles_outbreak_at_disney_anti_mmr_vaccine_activists_claim_disease_isn.html.

35. http://www.cdc.gov/mmwr/preview/mmwrhtml/mm6341a1.htm.

36. "Measles." World Health Organization fact sheet, updated March 2016. Available online at http://www.who.int/mediacentre/factsheets/fs286/en/.

37. http://www.ncbi.nlm.nih.gov/pubmed/9521945.

38. http://abcnews.go.com/US/oregon-faith-healer-parents-probation-sons-death/story?id=17273845; http://seattletimes.com/html/localnews/2017474241_apwafaithhealingdeath.html; http://www.nytimes.com/2004/11/16/health/16brod.html?pagewanted=print&position.

39. In the March 2015 issue of *National Geographic*, Joel Achenbach explains that tribal affinities create a strong emotional desire to espouse the same facts as other members of your tribe. Statements of belief about facts thus become identifiers of which group you belong to. For most people since the beginning of human culture, the desire to fit in has overpowered the desire to know the truth. As a dramatic example, consider those people who say they believe that climate change is caused by humans and those who say they do not believe it. http://ngm.nationalgeographic.com/2015/03/science-doubters/achenbach-text.

40. In *Why Truth Matters* (New York: Continuum, 2006, ch. 8), Ophelia Benson and Jeremy Stangroom point out that "inquiry, curiosity, interest, investigation, explanation-seeking, are hugely important components of human happiness." http://www.butterfliesandwheels.org/books/why-truth-matters/extracts/.

41. Jennifer Michael Hecht, *Doubt, a History: The Great Doubters and Their Legacy of Innovation from Socrates and Jesus to Thomas Jefferson and Emily Dickinson* (New York: HarperOne, 2004).

42. See Chris Mooney, "The Science of Why We Don't Believe Science," 2011; http://www.motherjones.com/politics/2011/03/denial-science-chris-mooney.

43. See Ziva Kunda, "The Case for Motivated Reasoning," 1990; http://www.ncbi.nlm.nih.gov/pubmed/2270237.

44. In response to the 9/11 "Truther" movement in general and BYU professor Steven Jones in particular. Quoted in the *Deseret News,* September 11, 2006. http://www.deseretnews.com/article/645200098/Controversy-dogs-Ys-Jones.html?pg=all.

45. Steven Pinker characterizes and defends the scientific way of knowing in "Science Is Not Your Enemy: An impassioned Plea to Neglected Novelists, Embattled Professors, and Tenure-less Historians": http://www.newrepublic.com/article/114127/science-not-enemy-humanities#.

46. For example, in *The Better Angels of Our Nature,* Steven Pinker masterfully merges the methods of science and the methods of history to draw scientifically supported conclusions from historical records.

47. Edward O. Wilson, *The Meaning of Human Existence.*

48. This point is explained in detail by Sam Harris in *The End of Faith* (New York: W. W. Norton, 2004).

49. Clay Farris Naff agrees with this point: "Terrorists are bad seeds. We may not prevent them all from germinating, but as humanists we can reduce the supply of fertilizer. We can do this in part by continuing to spread reason and science to neutralize supernatural belief." http://thehumanist.com/voices/the_humanevangelist/the-humanevangelist-how-to-keep-bad-terrorist-seeds-from-growing-hold-the-fertilizer.

50. In *Life After Faith: The Case for Secular Humanism* (2014), Philip Kitcher refers to this project as "the ethical project."

51. Russell Blackford offers the following comments on our concept of evidism:

Haley and McGowan are using the word "evidism" in a number of ways:

(1) Openness and responsiveness to evidence, rather than being dogmatic or taking claims on faith.

(2) Deference to the findings of science that are so basic and well-evidenced that they can be considered robust.

(3) Deference to "the scientific way of knowing" along with rejection of certain other alleged methods of obtaining knowledge, such as reliance on claims in holy books. This is an epistemological position.

(4) A refusal to believe certain supernatural and pseudoscientific claims that lack strong evidence, and thus a thesis about what kinds of entities, forces, etc., have a high probability of existing. This is an ontological claim.

(5) Philosophical naturalism, interpreted in a strong sense, involving disbelief in leprechauns, fire-breathing dragons, astral forces, ghosts, gods, and objective moral truths.

> (1)–(5) are not all the same thing. Evidism seems to be a combination of all these things, but in some situations they might come apart. (3) is perhaps the key point, as the others may follow from it, but the emphasis in this book is on (2). As used by Haley and McGowan, "evidism" is another word for "rationalism" or "scientific rationalism." "Rationalism" has a more technical sense in philosophy, where it is contrasted with "empiricism." But the more commonly understood concept of "rationalism" is the same as—or similar to—"evidism" as used by Haley and McGowan. Sometimes the expression "scientific rationalism" is used for this concept and this helps distinguish it from "rationalism" in the technical philosophical sense.

52. Novel by William Golding, 1954.

53. For examples, see http://secularhumanism.org/index.php/12 and http://instituteforscienceandhumanvalues.com/articles/neo-humanist-statement.htm.

54. Steven Pinker, *The Better Angels of Our Nature*.

# 2

# Challenges for Effective Communication

Before we dig into the new ideas to be shared, let us look at some of the challenges that frequently interfere with effective communication on topics like religion and politics and how to finesse them. If you already know this stuff or just want to skip to the meat, jump to chapter 3.

## Adjust your message.

Most of us are surrounded by friends who think like us, who reinforce our choices, who nod and laugh at the same things, who put us at ease. Most of us consume media of all kinds that reinforce what we already hold true rather than challenging us to consider other possibilities.

It used to take a bit more effort. For example, many radio stations in the late twentieth century had broad pop formats. The listener would stumble across unfamiliar things all the time—ska, reggae, punk, funk, new wave, blues, heavy metal, alternative rock, novelty songs—once in a while finding something both appealing and

51

entirely different from the songs he or she usually listened to. It was a broadening experience. Beginning in the 2010s, radio began to carve out narrow, carefully defined demographic slices. You're age 18–24 and like dubstep music? Great, I have the station for you. I promise you'll never have to hear anything else. No risk of encountering anything really new.

The website Pandora.com took this to a whole new level. Start with a song or artist you like, and Pandora will play songs or artists that are like the one you selected as a starting point. As each song plays, you tell Pandora whether it guessed well or not. Over time, the site learns your preferences so well that you never need fear hearing anything genuinely different.

The same is true in politics, religion, and social opinion. You can now find entire TV networks, magazines, talk radio programs, social media sites, and blogs devoted to reinforcing your opinions and protecting you from any serious risk of developing new ones. And all the while, the science of "behavioral marketing" sniffs behind you, studying what you do so those outlets can profitably feed you more of the same.

As a result, we are increasingly dividing ourselves up into smug, self-satisfied silos, each with everything it needs, including pundits devoted to telling us how smart we are to be in the silo we have chosen This increasingly thorough siloing not only shuts us off from our own growth but erodes our ability to communicate with or understand those outside of our own tightly defined cubicles. Churches do this with tremendous efficiency, but thanks in large part to the Internet, the nonreligious are also forming siloed communities—with the same good and bad results. Sometimes these religious or nonreligious communities devote themselves to good things like service and social justice, and sometimes they focus and facilitate a level of hatred and division that would not be possible without the reinforcement of that like-minded community.

We can't wish away the effect of siloing in our culture. But if we want to effectively share science and secularism across lines of difference, we have to start with a clear recognition of the fact of the silos so we can prepare our audience to hear what we have to say.

## Prepare the audience.

You might hear Neil deGrasse Tyson's point about cultivating a "sensitivity to the state of mind" of your audience knocking at the back of your mind right now. His gentle rebuke of Dawkins was basically a way of saying, "Richard, you have to recognize that some people are hearing you from inside very different silos! They don't necessarily see the world as we do. We have to start by being sensitive to that and preparing the audience for what we have to say."

Being right is not enough. To make a difference, we also have to be *heard*.

Suppose it's dinnertime at the family reunion, and your uncle starts in on something that has always made him furious: people on welfare. "I've been at the same job for thirty-two years, and you never saw ME asking anybody for a handout. But I look at every paycheck and see this huge handout that I'm giving to the government so they can pay people to NOT work! They sneak into this country illegally, then sit at home living the high life and shooting up drugs on my dime. I've really had enough."

As it turns out, you've recently done some research on this for a college paper, so you know he's wrong on every point. You're the one who's had enough! You formulate a reply in your head: *Seriously, do you ever think before you speak? You're so wrong I don't even know where to start! Most assistance programs require people to work in order to receive the benefits, did you know that? And food stamps average only $1.50 per person per meal, which is hardly the high life.[1] That wouldn't even pay for the enormous hamburger you're about to put into your face. And exactly ZERO illegal immigrants are on welfare, by the way. They aren't eligible for any government assistance except emergency Medicaid.[2] As for drugs, people on welfare are less likely to be on drugs than the general population![3] Game, set, and match!*

In the end, you say nothing. You'll just upset the nice family dinner by yelling back at him, you decide, and you're not going to change his mind. Well, if you read that script, you're right on both counts. You would have left him no option but to defend himself, angrily. Everyone will be embarrassed, and no minds will be changed, with the possible exception of people thinking less of you for being disrespectful, *even if you're right.*

But is yelling the only option, and is changing his mind the only goal?

Look around the table. Even if Uncle Brian doesn't change his mind, your whole family—including several impressionable kids—just heard that lazy immigrant drug fiends are living off the taxes of hard-working Americans. So point #1: there is almost *always* another audience. Whether posting on Facebook or having a conversation at a party, what you say influences not just the person you are talking to, but everyone listening in.

And I wouldn't even give up on Uncle Brian just yet. Yes, if you attack the way you imagined, the result will be poor. So before you pour out every fact you know, prepare your audience. Empathize with him. "I know what you mean about the taxes. I remember the first time I saw that on my paycheck: I just about fainted! And I wondered the same thing: is this going in somebody's pocket who hasn't earned it?" Even if you've never wondered that in your life, find a way to say *I understand how you feel.* "So I did a little research, and I was relieved to find out that most people on welfare actually work, and a lot of them work full-time but just have trouble making ends meet."

That's it. Don't move on to the immigrants and drugs and high life or you might tip the balance into embarrassment and anger again. You empathized, you defused *one* misconception, and you presented it as good news. You also sent the kids around the table a subtle message: Uncle Brian is sometimes mistaken. And you did it in a generous way that didn't undercut your own reputation. In fact, people will be all the more likely to listen to you the next time an issue comes up.

This approach is related to a technique called Nonviolent Communication, a powerful and effective concept developed by Marshall Rosenberg and others. It always starts with **empathy**—making an effort to grasp and feel what the other person feels, to hear things as s/he hears them, and to frame what you have to say accordingly. It can leave people feeling heard, understood, and honored—*even if they continue to disagree.* It can lead to amazing breakthroughs by recognizing that win and lose are not the only meaningful terms in dialogue.

We came across an example of this kind of communication in an unexpected place: a children's song called "My Brother the Ape" by the band They Might Be Giants.[4] The clever, catchy song was written to introduce young people to a specific implication of evolution—the challenging idea that all living things are related.

This idea has to get past one of the most ingrained human conceits—that we are special and separate from all other living things. So rather than simply bang on about the fact that we're all related so get over it, the song shows a "sensitivity to the state of mind" of the little humans who will be listening. "My Brother the Ape" is sung from the perspective of someone who has trouble letting go of human specialness and accepting his kinship with all living things. Indeed, for someone who has been raised with the notion that humans are specially created in the image of God to "rule over the fish of the sea and the birds of the air, over the livestock, over all the earth, and over all the creatures that move along the ground" (Gen 1:26)—or even coming from a pretty natural position of human chauvinism—evolution represents a serious demotion and a choking slice of humble pie.

A song that empathizes a bit with that reluctance can offer a place for the listener to stand, and sing, while they consider whether or not to join the evolution. It is a brilliant approach and an effective example of sharing reality.

## Focus on marketing issues.

Most people think of marketing in the narrow context of commercial advertising. A company markets its products and services to you, the consumer, by trying to convince you that they are desirable enough to pay for—that your life will be better with them than without them. But effective marketing methods can be used much more broadly. Police offering gun buyback programs are marketing a plan to reduce violence. A politician asking for your vote is marketing her candidacy to you in the hope of securing your vote.

Ideas can also be marketed. If you think an idea or belief I hold presents risks for your well-being or the well-being of your community or the world at large, you can try to convince me to accept another idea in its place, one that you find less risky. If I accept your idea, our community is that little bit better and safer.

Key to marketing is how we define and use words. Without words, we can't understand important concepts. To make concepts easy to understand, we need to select and use effective words. Some of the following suggestions urge careful use of words relating to reality. Other suggestions urge spreading of specific concepts that can best be accomplished with a focus on marketing issues.

The remainder of this book is structured around twenty-five such concepts, all easy to understand and all important concepts to consider when sharing reality with your friends and the world.

## NOTES

1.  Welfare Information at http://www.welfareinfo.org. Accessed March 12, 2015.

2.  "10 Myths about Immigration," Teaching Tolerance. Available at http://www.tolerance.org/immigration-myths. Accessed March 12, 2015.

3.  "The Myth of Welfare and Drug Use," *Daily Beast,* August 20, 2013. Available at http://www.thedailybeast.com/articles/2013/08/30/the-myth-of-welfare-and-drug-use.html. Accessed March 12, 2015.

4.  You can hear in its entirety at www.youtube.com/watch?v=cQ_WeLi09p0.

# 3

# How to Spread Secularism and Science

With the background material now established, let's get into the new ideas and practical advice. We start by presenting sixteen suggestions for how you can help your friends and loved ones reduce conflict and harm in their lives by accepting the scientific consensus on the facts and approaching their interactions with others from a secular position. Each suggestion also applies to how leaders of secularism and science can spread their messages through more effective marketing to all people, including those in the less culturally evolved parts of the Middle East and Africa.

## Explain that traditional ways of knowing, including religions, are natural and often wrong.

The world has been dangerous and violent for most humans in most places and times. As pre-humans developed, the brain evolved intuitive ways of assembling and retaining beliefs about everything relevant to making decisions that affected survival or thriving. Before the advent of language, using individual intuition, each person assembled his or her own set of beliefs on which to base decisions. Such beliefs would surely have varied considerably from person to person.

The intuitive ways of knowing that evolved naturally over hundreds of thousands of years work well enough in such creative fields as visual arts, music, dance, and storytelling. Unfortunately, evolution also gave our brains intuitions that, if unchecked by education, lead us naturally to believe "facts" that aren't true.

The leading theory among anthropologists is that, more than 10,000 years ago, humans believed that spirits caused all events, including human actions. Evolution caused us to interpret undecipherable sounds or glimpses as likely an agent, such as an attacking person or animal or spirit, intending to cause us harm. While these interpretations were often false positives because there was nothing harmful approaching, the false positives caused us no harm, whereas a single false negative would mean the end of our gene line.[1] These intuitive ways of knowing that saw spirit agents in movement everywhere then evolved naturally over hundreds of thousands of years in separate cultures to create various gods, spirits, and countless mutually contradictory religions.[2]

Naturally evolved ways of knowing generated many other false factual beliefs, only some of which were ever formally captured in organized religion. These include beliefs in omens, ghosts, spirits, witches, fate, destiny, providence, karma, charms, spells, curses, astrology, tarot cards, homeopathy,[3] water-witching, lucky and unlucky numbers or objects or events, unwarranted fear of vaccines, unwarranted fear of fluoridation of water supplies, beliefs about special days such as Day of the Dead, and the supposed effects of amulets and talismans.

As early humans began to develop language, beliefs could be passed on by words from one person to another, particularly from older to younger. Evolution has seen to it that children are credulous. There is simply no other way to learn everything they need to know in order to change from helpless, grasping newborns into fully functioning adults in about 6,000 days. That is good in one way. When we are children, the tendency to believe it when we are told that fire is dangerous and cliffs are not to be dangled from helps us, in the words of Richard Dawkins, "to pack, with extraordinary rapidity, our skulls full of the wisdom of our parents and our ancestors"[4] in order to accomplish the complex feat of becoming adults.

If children were to examine and question everything they are told, they would never make it to adulthood. Some of this passed-on information consists of tested truths based on good evidence. Unfortunately, untested "facts" based on intuition, inspiration, tradition, dogma, superstition, and religion are also passed down, often with great conviction.[5]

From the perspective of educated people today, many naturally evolved forms of "knowledge," both religious and nonreligious, appear clearly to be illusory and ill-founded. But an educated perspective should also include empathy for those who have believed in ill-founded things, and even those who continue to do so. Before the development of scientific methods, these were the only available sources of knowledge on many topics. Given a choice between seeing ourselves as powerless victims of an unfeeling universe or as the beloved children of an all-powerful deity that we can ultimately influence through ritual and prayer, it is easy to see why humans would favor the latter.

It is easy and common to suggest that, in the absence of a reliable means of knowing, we should simply say "I don't know" instead of filling that space with false knowledge. But for most of human history, nothing would have felt more vulnerable than an empty hole where knowledge should be. Intuitive false knowledge would rush in to fill that vacuum. And without reliable ways of discerning the truth, the most common human response would be to choose beliefs that favor survival or match our preferences.

You might say that this was all well and good when we didn't know what made the sun come up or what caused diseases, but now we know these things, so it is time to leave our fearful beliefs behind. The problem with this is twofold: many of those beliefs were tailored precisely to the satisfaction of human instincts and, despite advances of scientific understanding, reality is still difficult to accept. We are still vulnerable; there is still much we do not know and cannot control.

The choice between fantasy solutions and harsh reality can make the truth a hard sell. The benefits of wealth, technology, health care, and education protect many of us from the paralyzing fears that can lead to accepting false conclusions about the world. But many others aren't so lucky. And once they accept false ideas, they pass them on to future generations.

Science as a formalized way of knowing is only about 500 years old, while the prescientific, naturally evolved, intuitive ways of knowing "facts" have dominated human culture since the beginning of culture itself and continue to do so in all cultures today. That is a weighty tradition to overthrow.

Recognizing this, and feeling empathy for those less privileged in the struggle to overthrow that weight, is not just about being "nice." It is a vital part of effectively sharing reality. And, because false beliefs often cause harm, it is a task worth doing effectively.

## TEACH THE THREE MINIMAL ELEMENTS OF SCIENTIFIC UNDERSTANDING: (1) SCIENCE IS RELIABLE, (2) ANYONE CAN LOOK UP WHAT IS KNOWN, AND (3) NOTHING IS KNOWN FOR CERTAIN.

To accept the scientific consensus on the facts (evidism) does not require an understanding of scientific methods or any ability to practice them. Acceptance can be taught by rote, such as by parents or teachers. Here are the three minimal elements of scientific understanding that should be taught to everyone.[6]

### (1) SCIENCE IS RELIABLE BECAUSE IT IS VETTED, AND (2) YOU CAN LOOK UP THE ANSWERS.

Some science-minded readers will gasp at this. *Look it up?!* Science is about rigorous, repeatable experimentation, careful observation, and presenting one's findings to competent peers for review and critique! You're right, of course. That's exactly what the practice of science is about. But most people do not need to understand this or be able to practice it.

Most of the knowledge you and I have about the world and universe around us was not gained by first-hand experimentation and peer review. It was gained by accepting the findings of others who demonstrated their own respect for and adherence to the methods of science by putting it through the gauntlet of rigorous peer review. Consider your opinion that the earth orbits the sun rather than the other way around. How did you come to that conclusion? Not by observation, since the apparent "rising" and "setting" of the sun each day clearly suggests the opposite of the truth. You, like most people, learned that the earth orbits the sun from parents and teachers who in turn were delivering the findings of scientists over the centuries who did the hard first-hand work of figuring out this exceedingly nonintuitive truth.

The first and most important step in accepting the scientific consensus on facts (evidism) is understanding that the facts determined by science have been developed by the collective efforts of large numbers of people who have checked and criticized one another's work to achieve a reliable consensus. Science has a system for weeding out what is false. This makes the scientific consensus on facts worthy of reliance by everyone.[7] But the longer the chain of intermediaries between you and the science—in other words, the longer the game of Telephone that takes place in passing the information down to you—the greater the chance of introducing errors and misinformation.

Fortunately, we live in an age of unprecedented access to information via the Internet. It is hard for many people to recall a time before Google made its stunning mission to "to organize the world's information and make it universally accessible and useful" a practical reality.

The second important element to teach about science is that anyone with an Internet connection can look up the scientific consensus answer to nearly any factual question using resources such as Wikipedia[8] or the simplified Wikipedia pages.[9] You don't need a strong scientific education to do this, so there is no longer an excuse for being a science denier.

### (3) THERE IS NO 100 PERCENT CERTAINTY FOR ANY FACT.

The third element for accepting the scientific consensus on the facts (evidism) is understanding that each "fact" has an associated probability of being correct, and no facts are known to 100 percent certainty.[10] Human knowledge is a constantly improving fuzzy approximation of reality. Each "fact" reflects a "model" or "theory" of reality, and each of our models of reality might turn out to be not quite right. For example, until Einstein improved our understanding of physics, it was an established "fact" that there is no limit to the speed that a moving object can achieve. Now our current model of reality has a firm limit equal to the speed of light.

Well, what about simpler and more obvious facts—that your body exists, for example, and that you are reading this book? It is fair to be extremely confident in these things, but can you really say you are absolutely, 100 percent certain? The existence of your body and your reading of this book might be taking place in a dream. That is not at all likely. The chances that your body exists and you are reading this book are very, very high—but they are still not 100 percent.

Thus, for people who accept the scientific way of knowing (evidism), the words "true" and "false" are not absolute as they seem but merely shorthand for particular probabilities. We accept as "true" those facts that have a very high probability of being true—likely enough that we rely on them—and we dismiss as "false" those proposed facts that have a low probability of being true.[11] For proposed facts with a middling probability of being true, we might take no position or we might make branching decision plans with branches based on each possibility. These are the cases of "maybe."

For example, there is a very small but larger than zero probability that leprechauns exist. If you happen to believe in leprechauns, to get an idea of how small the probability is, substitute "Flying Spaghetti Monster." Both are equally unlikely. For any proposed "fact" with such a small probability, we can describe it as "no more likely than leprechauns" and the shorthand is "false."

Of course, science does not have an answer for all factual questions, and there are areas of factual inquiry where the present best answers are tentative and of moderate probability. These are areas where there is honest difference of opinion and we assign middling probabilities to each possible answer (the realm of "maybe"). However, in the last one hundred years, science has reached the point of having clear answers to nearly all factual issues relevant to understanding the reality of human experience.

It is important for everyone to understand that nothing is known to 100 percent certainty. Failure to understand this causes endless failures of communication on the subject of knowledge and great difficulty for students who just want answers. When we say we "know" something or that something is certain, we are speaking in shorthand. What we mean is that it is almost certainly true, likely enough true that we rely on it, not that it is impossible for it to be untrue. In our shorthand, anything that is no more likely than leprechauns is said to be "false" even if it might in fact be true.

### SCIENCE HAS ESTABLISHED BASIC FACTS.

This book is not intended to persuade anyone of what the facts are. If you doubt any of the following basic facts, read the books cited in the notes. To date, insufficient factual evidence has been found to support anything beyond a tiny probability of any of the following concepts being real. While there is a very small chance that each exists, each is no more likely than leprechauns:

- a god[12]
- a creator
- spirits that cause events but cannot be seen
- spirits of your ancestors who can help you navigate life[13]
- a soul that survives death
- an afterlife
- reincarnation
- a mind or thoughts separate from the physical brain
- astrology

- homeopathy

- faith healing

- extra-sensory perception

- a valid reason for ordinary people to reject standard vaccinations

- a valid reason to reject nonexcessive fluoridation of water supplies

- mental telepathy

- water-witching

- omens

- lucky or unlucky days or numbers or talismans

- any objective source of human values or morality outside of humans.[14]

In the scientific way of knowing (evidism), we don't say we know for certain that such things do not exist. We merely recognize that the evidence for them is no stronger than the evidence for leprechauns. While the probability each of these things exists is not zero, the chances of existence are trivial, so small that we take no actions based on those chances. The probability is so small that we conclude, for all practical purposes, that they do not exist.

## TEACH PEOPLE TO SEPARATE FACTS FROM NONFACTS.

The scientific way of knowing facts is more clearly understood if people learn to separate all concepts into facts and nonfacts. Nonfacts encompass values, opinions, nonsense, etc.[15] By "facts" we mean scientifically objective facts, in contrast to socially constructed facts such as some of what counts as "facts" in a moral or legal system.

While science determines all objective facts and can be useful to help clarify values and opinions in particular situations, it cannot determine any values or other nonfacts. Values are inherently subjective. Acceptance of science to determine all facts allows each person to focus more clearly on tasks that are essential for the well-being of individuals and society—developing good values. One can simply look to science to get the answers on all the important facts, but there is no objective way to get answers on the best values. Developing good values requires learning and reflection. Most of the learning comes from the humanities, including religion.

Consider for example the questions, "When does human life begin, and when does it end?" People posing these questions are typically showing that they hold a value in which there is something special or valuable or meaningful about "human life," whatever that may mean to them. To separate out the facts, we start with the following observations.

Each living human body is composed of millions of cells. At the cellular level, the concepts of life and death are relatively clear. A living cell is one that has the ability to divide or to perform a function that aids the functioning of an organ. If a cell changes such that it can do neither of these things, it is dead. In any "living" organism, there are millions of living cells and millions of dead cells.

Every living cell was formed by the division of a pre-existing cell or by the merging of two pre-existing cells. Consequently, looking at the cellular level, it is meaningless to ask when a "life" began. Either each cell had its own day of origin when its life began by a merger of cells or division of cells, or the day of origin for all cells on Earth was the day that the first living cell evolved millions of years ago. Which you choose is a matter of your preferred definition, and neither choice sheds light on interesting questions.

At the level of a multi-celled organism, what is "life" and what is "death"? There is no determinate scientific answer to this question. At all times, some of the body's cells are alive and some are dead. No new human "life" has been created in the last million years; each individual collection of living cells is just the result of dividing and combining of cells over eons. There is no way to decide factually when an animal is "alive" without selecting a somewhat arbitrary definition such as "when the heart begins to beat," a definition that may not be suitable in some situations. For example, if we use the heart beating to define life vs. death, are people "dead" when their heart stops beating but an artificial pump moves their blood to keep the body and brain otherwise alive? When life begins and when it ends are therefore values questions, not fact questions; it depends on which aspect of life is valued for a particular life or death determination.

By contrast, consider the related question of whether humans have souls or karma as those words have been used in religious and spiritual traditions (as opposed to "soul" just meaning what a brain does or "karma" just meaning the effects of one's reputation in a community). The factual answer is no: the probability that there is anything that can be called a soul or karma in the spiritual sense is so low that it is no more likely than leprechauns. Any values that are based on humans having souls or karma are poorly founded and deserve to be reconsidered.

Now consider whether the killing of a person is murder. "Murder" is a value-laden word, meaning that the person who uses it is expressing a value, not an objective fact. If we separate out the facts, all we can say is the cessation of a human's heartbeat (or however we agree to define death) was caused by an action of one or more other humans. If the dead person was terminally ill and in great pain and asked to be put to death, then a person might decide, as a matter of values, that it was not "murder."

As another example, consider "ownership." If I say "my arm" or "my eyes," it is a statement of objective fact because the object is physically attached to me. But if I say "my property," it is a statement that society has decided, as a matter of values, that I should have certain rights vis-à-vis other humans to exert certain kinds of control over that property. There is no objective relationship between me and the property unless I am holding it or sitting on it. Although it may be considered to be a fact in a legal system, "my property" is a values statement, not a statement of objective fact.

It is difficult for people to learn to separate objective facts from nonfacts because, in human language, we use the verb "believe" to apply to both facts and nonfacts. For example, Stephen Hawking once stated his opinion on a question of fact using the word "believe": "I believe alien life is quite common in the universe, although intelligent life is less so. Some say it has yet to appear on planet Earth."[16] John D. Rockefeller used the same word in a statement of values: "I believe in the dignity of labor, whether with head or hand; that the world owes no man a living but that it owes every man an opportunity to make a living."[17] When people say, "I believe . . .," it is often unclear whether the person means facts or values.[18] It would help everyone's understanding of these issues if we had a verb that applies only to facts and we used a different verb for values and opinions. But we don't have such a helpful pair of verbs, and there is little chance such words will be added to our language in the foreseeable future. For now, we have to rely on follow-up questions to determine in which sense the word is meant. Not a very efficient way to proceed, but no language is perfect.

## ACCEPT THE TRUTH ABOUT ALL FACTS— CRITICAL THINKING, SECULARISM, AND SKEPTICISM ARE NOT ENOUGH.

Some people who do not accept the scientific consensus on facts think that statements of fact that aren't intuitively obvious are arbitrary. They accord equal weight to mutually inconsistent proposed beliefs about facts. For these "fact relativists," facts are relative to the culture of people: what people of one culture believe about facts can be in conflict with what people of another culture believe, and neither can be considered wrong. To those who accept the scientific way of knowing (evidism), this is profoundly erroneous and holds back human understanding and happiness.

When some people hear the proponents of various religions touting inconsistent facts, they know that all but one are wrong or that all are wrong, but they conclude that the truth is not important. Many of them say, "It doesn't matter which religion you choose, just pick a group and believe in their religion; you can't do without beliefs entirely, and you can't know which set of beliefs is right, so just pick the group that you like the best or that feels intuitively right to you." From this theistically relative perspective with respect to facts, believing in no gods is just another option that one might choose and has no better reason behind it than believing in one god or a thousand.

People can take such an approach to all kinds of facts where the truth is not intuitively obvious. For example, there are people, religious and nonreligious alike, who believe homeopathic medicines are effective beyond placebo effect. They seem to adopt this belief by following the beliefs of their friends or people they consider to be authorities. Scientific studies have conclusively demonstrated that homeopathic (in the sense of the word urged by its leading proponents[19]) remedies being effective by anything more than placebo effect is no more likely than the existence of leprechauns.

As a second example, there are people who identify as nonreligious but believe that each tooth in the mouth has a special connection to a different organ in the body, and that you can learn about the health status of the organ by examining the health status of the tooth (an Ayervedic belief).[20] Such beliefs can result in harm from people failing to avail themselves of scientifically proven remedies.

Tamara Sophie Lovett treated her seven-year-old son Ryan, severely infected with streptococcus, with homeopathic and herbal remedies rather than obtain from the Canadian healthcare system the usual remedy of penicillin, which is nearly always effective. When Ryan died in 2013, Ms. Lovett was appropriately charged with criminal negligence.[21]

We discussed above more tragic examples, including people dying or being permanently harmed from denial of vaccination science or from a faith-healing approach to appendicitis.

Leading people to critical thinking and secularism, while essential, is not enough to lead them away from a relativist view of truth about facts. If they will not accept the scientific consensus about connections between teeth and organs or homeopathic remedies, it will likely be difficult to persuade them to accept other conclusions of science.

It is not enough for evangelists of cultural evolution to lead people merely to secularism and away from god beliefs; it is more important to lead them also to accept the scientific consensus on all facts (evidism), not just facts relating to deities and spirits. Unless people come to understand and accept whatever scientific consensus has been established on all aspects of reality, they will not understand why an atheistic way of knowing is better than each of the theistic ways of knowing. The choice will continue to appear arbitrary to them.

The scientific way of knowing is more than just a secular and atheistic outlook. Homeopathy, astrology, and vaccine denial are generally considered to be secular rather than religious. However, science shows the chance of each of them being valid is no more likely than leprechauns and they are each just as incredible as any religious claims. Likewise, if you believe in omens or lucky or unlucky numbers or days, you can still be an atheist, but you have not fully accepted the scientific consensus on facts and are not an evidist. Such inconsistency can be harmful. By embracing unscientific beliefs, even with the best of intentions, you run the risk of adopting poor values because of your false beliefs about facts.

Unlike atheism and skepticism, the scientific way of knowing (evidism) is not merely a negation of invalid sources of knowledge; it is a positive view that affirms valid sources of knowledge on all factual topics. It exults in the prospect that there is no part of reality that cannot be discovered through inquiry consistent with science.

The scientific way of knowing isn't natural or intuitive for humans. Evolution did not wire our brains to know intuitively that some authorities are wrong, or that hundreds of years of testing, retesting, and sharing observations and conclusions will yield results far more accurate than our intuition, or that nothing can be known with 100 percent certainty. Education in the scientific way of knowing is needed before a person can overcome the siren song of intuition and fully accept the scientific consensus. Those who lack that education can, with the best of intentions, refuse to vaccinate their children or vote against nonexcessive fluoridation of city water supplies or, without adequate reason, take other actions that are harmful to the interests of others. They can even do these things while being secularists and atheists.

It is not enough to be a secularist and a skeptic and an atheist; it is even more important to also be an evidist, for the sake of other people in your society and for your own happiness. People who become evidists are certain that it makes them happier.[22] As illustrated in figure 5, data from the 2013 UN World Happiness Report and data

## FIGURE 5. HAPPINESS VS. RELIGIOSITY BY COUNTRY

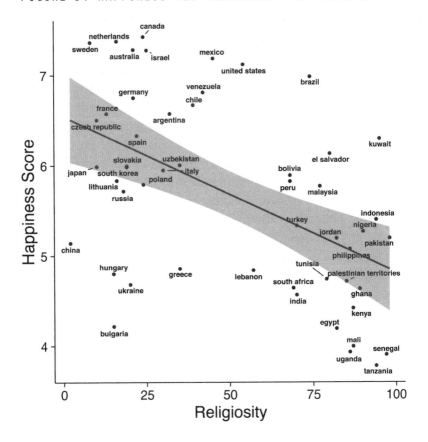

NOTE: PEARSON CORRELATION: -0.52

SOURCE: GRAPH CREATED BY GLUON SPRING, USED WITH PERMISSION

from various Pew Research Center surveys on global attitudes show that people in less religious countries report they are happier.[23]

## A FALSE UNDERSTANDING OF FACTS IMPEDES DEVELOPMENT OF GOOD VALUES.

Traditional religions claim that morality is based on the "facts" that a god exists or that people have souls or karma. Once these "facts" are understood to be false, what is the basis for morality? The scientific way of knowing gives no answers to moral questions or any other issues of values other than questions like which values cause more conflict and which cause less. It only provides all the facts.

That doesn't mean facts are entirely unrelated to values. As the pioneering sexuality researcher William Masters put it, "Science by itself has no moral dimension. But it does seek to establish truth. And upon this truth, morality can be built."[24]

To understand the connection between values and facts, consider the following values questions.

Is it acceptable for women to be treated as the intellectual inferiors of men? After centuries of considering such unequal treatment to be justified, it is now the overwhelming consensus in most of the world that such treatment is wrong. Why the change? Because from every direction of scientific research, women are the proven intellectual equals of men. The facts do not provide the values directly, but by starting with what we know, we can develop good values.

Is it acceptable for rights that adults enjoy to be withheld from ten-year-old children? Your next question might reasonably be, "Which rights do you have in mind?" You would ask this in order to run the decision through a filter of known facts about children's abilities compared to those of adults. You might decide that a ten-year-old lacks the needed maturity and physical abilities to drive a car, but you might allow a child to choose his or her own clothes in the morning or to ask for the reasons behind parental decisions, both of which are freedoms recommended by child development experts based on known research outcomes. Once again, the facts don't dictate the value, but they lay the groundwork.

In the same way, a false understanding of facts impedes development of good values. A parent who values corporal punishment as a method of discipline, believing it will lead to a strong character for the child, is holding a value that directly contradicts what we know to be true—that corporal punishment leads to eleven measurable outcomes, ten of which are negative for the child. These negative outcomes include a greater likelihood that the child will eventually suffer from depression, exhibit violent tendencies, have an impaired relationship with the parent, and abuse his or her own children.[25]

Many people who once thought their values were based on a god and then came to understand and accept the scientific way of knowing, adopt values statements of humanism[26] or Unitarian-Universalism[27] as a reflection of their own values. Others invent their own humanistic values as an adjunct to the intuitive values provided by evolution of the human brain. When the scientific way of knowing is accepted, established values that were erroneously based on wrong "facts" fade and are replaced with other values based on a more accurate understanding of facts. The quantity of values that each person holds is not reduced by acceptance of the scientific way of knowing; the values are merely adjusted.

The scientific way of knowing facilitates good decisions on the important issues of values, including morality and ethics, because people no longer base these decisions on answers they believe are provided by a deity or some other source outside of humanity.[28] For example, when one comes to understand that there is no scientific basis to treat human conception as a dividing line between a new "life" and pre-existing viable living gametes, or to believe that any human has a soul, it becomes easier to adopt the value that each pregnant woman should be allowed to decide whether to abort her pregnancy. The facts don't dictate the value, but they provide the precursor for good values formation.

Accepting the scientific consensus on the facts and holding no inconsistent values also help reduce suffering from disease. People who deny the science of fluoride and successfully oppose fluoridation of public water supplies—a position with no basis in science provided the fluoridation levels are not excessive—increase the risk of tooth decay for everyone in the community. People who deny the science of vaccination increase the presence of preventable diseases for themselves and for others in their communities and the world.

## BE COCKSURE ABOUT ESTABLISHED FACTS AND CAUTIOUS ABOUT VALUES.

The philosopher Bertrand Russell once said, "The fundamental cause of trouble in the world today is that the stupid are cocksure while the intelligent are full of doubt."[29] Because you are probably confident of your membership in the latter group (though hopefully not cocksure), this quotation probably hits you with the force of revelation. We have all known people whose political and social opinions are matched with an obnoxious, strutting certainty completely unjustified by the well-established facts. At times, their arrogance seems to increase with their ignorance. This is such a common occurrence that it has a name—the Dunning-Kruger effect, named for the two researchers who first described it in 1999.[30] The less competent a person is, the less he recognizes his lack of competence and the more confidence he therefore has in his opinions. It is the scientific confirmation of Russell's pithy remark.

Given all that, you are justified in wanting to stay miles away from cocksureness. But it is important to recognize that confidence itself isn't the problem, only *unjustified* confidence.

Religious people are often arrogantly certain about some "facts"—the age of the earth and the relationships between species, for example—that science has demonstrated are entirely wrong. Their arrogance is unjustified. However, on facts that have been vetted by scientific methods and established with a robust consensus, everyone can be rightly confident, even cocksure. Where people rely on facts with middling probabilities, they should be humble and respectful when others disagree.

But be cautious when arguing the strength of your position based on values. Unlike facts, there is no objective way to establish which values are right or wrong. The values that are a part of evolved human nature are not right merely because they are intuitive for humans. Just because murdering people of other tribes or keeping slaves is part of evolved human nature doesn't make those practices right. Even if something was once highly adaptive for human evolution, that has no bearing whatsoever on the values question. As human culture has evolved, a consensus has emerged that certain values like those above are wrong even though they are natural. Once all cultures on Earth agree on a value, this makes it appropriate to be highly confident, even arrogant, that the value is good for human culture.

Consider the value that some have adopted that it is wrong to kill animals for human food. Many people are arrogant vegetarians. However, vegetarianism is inconsistent with evolved human nature, and there is wide disagreement with this value. Consequently, arrogance is unjustified. Neither those who hold this value nor those who oppose it have an adequate basis for arrogance. The question should be approached from both sides with respect for the value choices made by others on this topic.

## BE WARY OF VALUES THAT PERPETUATE UNWORTHY ASPECTS OF HUMAN NATURE.

As we are social animals that succeed through cooperation, evolution gave us an intuitive moral sense (intuitive values) about interacting with members of our tribe. Joshua Greene argues in *Moral Tribes* that we should mostly trust our intuitive values about interacting with our families and friends and group members,[31] but when it comes to interacting with people of other tribes, races, or languages, our intuitions often lead us to animosity and conflict. The reason is simple: we are naturally inclined to treat those closest to us with the maximum amount of empathy and compassion and are naturally inclined to treat those further from us with distrust and suspicion. During the Paleolithic Era, when humans walked the African savanna in small kin groups, this made excellent sense: strangers from another nomadic band put more strain on local resources and were therefore a threat to one's own band's survival. But we now live in a very different world, one in which the interests of very different people are densely intertwined. It is now counterproductive to treat those different from us as a threat or to act against their well-being. We can now cognitively appreciate that we are no more special than they are, and there is no justification for us to treat them differently if we expect them not to treat us differently. The flourishing of all humanity rather than merely one tribe is now a worthwhile goal, so any values that lead us to tribal conflict rather than cooperation in the present day are candidates for rejection even if they are natural.

Values are passed down across generations by tradition. We should be suspicious of values passed on by traditions that have not been carefully examined. Values based on human traditions are typically infused with tribalism, racism, and xenophobia—natural tendencies that are nonetheless undesirable.

In addition, males have natural tendencies toward violence, rape of females, and sexism. The biologist David Lahti puts it bluntly: "From an evolutionary perspective, considering other social species on this earth, it is remarkable that a bunch of unrelated adult males can sit on a plane together for seven hours in the presence of fertile females, with everyone arriving alive and unharmed at the end of it."[32] It only happens because we have evolved culturally at a fast enough rate to overcome our natural tendencies. And thank goodness for that.

Intuitive and traditional values are likely to make greater allowance for these outdated tendencies than may be best for human culture.

To summarize, generally trust your instinctive values toward your friends, family, and groups, but suspect your instinctive values toward:

- females and other males if you are male; and
- members of other national, cultural, racial, religious, or tribal groups.

The more advanced human cultures recognize that intuitive values driven by these instinctive, natural tendencies are not the best values for cultural progress. It requires education and conscious thought to develop values that overcome these natural tendencies.

## TEACH THAT NATURAL HUMAN TENDENCIES TOWARD TRIBALISM SHOULD BE TRANSCENDED OR CHANNELED APPROPRIATELY.

Because values are inherently subjective and not provable by science, this book takes no strong position on any values except for the value of accepting all the conclusions of science and the three values essential to secularism listed above, the first of which requires suppression of culturally harmful tribalism. Tribalism is the tendency to identify and empathize with those most similar to ourselves, and to shun, distrust, and even actively oppose those who are less similar. Whether or not tribalism was adaptive in the distant past, it is clear that most such thinking is overwhelmingly contrary to human flourishing in the present. So tribalistic thinking, emotions, and actions, including those based on race, religion, or ethnicity, should be broadly discouraged for the general good. Although this value is not *objectively* right, persuading everyone to transcend this tendency is essential to reducing conflict. There is no acceptable alternative.

That is not to say there aren't *unacceptable* alternatives. If a single racial and religious group were to assert dictatorial authority, ruling all others with strict and effective policing, that could reduce tribal, racial, and religious conflict. This option has been tried, and more than once. It was the basis of the peace imposed on tribes conquered by the Roman Empire, the famous *Pax Romana*. After centuries of continuous warfare, Rome ruled most of Europe and the Mediterranean for over 200 years (27 B.C.E.–192 C.E.), allowing most of its citizens to live in peace without engaging in military conflict with other tribes. But they did so only by the often brutal suppression of liberation movements and uprisings among their subject peoples. Eighteen hundred years later, the architects of National Socialism in Germany envisioned a similar imperial "peace" to follow the Second World War.

This method of achieving peace will never be acceptable to people who are not members of the ruling tribe or religion, will result in widespread harm and dysfunction, and will never be strong enough to suppress all rebellion. For this reason, we can reject it as a viable alternative. With no other alternative, it is clear that transcending most of our natural tribalistic tendencies is desirable and that convincing others to embrace that value is a worthwhile goal. "Secularism is and always will be the least bad alternative for achieving peace in complex, religiously pluralistic societies."[33]

Even among those who embrace this value generally, there is some debate about how to implement it. Some suggest that we can never entirely transcend the powerful urge toward tribalism, and that we should instead sublimate our tribalistic tendencies into regulated competitions such as team sports with fanatical fans. They argue that such an outlet is not only harmless and enjoyable, but can even act as a safety valve, reducing the chance that these tendencies will erupt in more violent and inhumane ways. It is a credible hypothesis, but one that is hard to test.

Of course, even sports tribalism is not always nonviolent. The first recorded sports riots occurred in the year 531, in Constantinople, the heart of the Byzantine Empire, when rioting fans after a chariot race caused many deaths. To quell the riots and regain control, the emperor convened another race, colluded with the fans of one team to suddenly leave the stadium, and then slaughtered the remaining 30,000 fans of the other team. Sports events still lead to violence today. U.S. fans routinely overturn cars and set fires even after *winning* championship matches in any number of sports, and soccer hooliganism has claimed hundreds of lives worldwide in the past decade alone. One such riot in Egypt in 2013 claimed 74 lives.

There are other ways to tap and channel tribal identity with even less residue of conflict and competition than sports. When a church or school or humanist organization rallies its members to donate canned food to the poor or volunteer at the scene of a natural disaster, they often do so by appealing directly to that shared tribal identity. They don't generally call it "tribal"; but when they say, "We are called as Christians to help those who are less fortunate," or "Show your Tiger spirit! Donate to the canned food drive," or "The highest aspiration of humanism is to care for each other," each group is attempting to channel our innate tribalism to a positive end.

## Urge each member of a religious congregation to accept the scientific consensus on facts.

Religions flourish because they serve human needs—for community, for identity, for guidance, for a sense of purpose, to focus a sense of personal wonder and gratitude, and for a source of rapture.[34] None of these needs require false claims of fact, and all can be fulfilled by social organizations, including religions, without the slightest mention of such claims. We should encourage each religion to get out of the business of spouting bogus facts and instead focus on promoting good values, offering community, offering identity, offering rapture through practices such as music, dance, yoga, and meditation, and doing whatever can honestly be done to help people feel they have a purpose.[35]

But encouraging religions to change does not have a long and glorious history of success. Stasis is woven into the very fabric of religious institutions. That it took 359 years for the Catholic Church to apologize for the arrest and censure of Galileo for stating publically that the earth revolves around the sun is only the best-known example of the pace taken by change in most religions.

The Vatican's achingly slow movement toward acceptance of evolution by natural selection is an even better example of the difficulty. Since Darwin published *On the Origin of Species* in 1859, a few popes have tinkered at the margins of the question. They rarely mentioned evolution at all during the nineteenth century, unless to affirm repeatedly "the special creation of man." In *Providentissimus Deus* (1893), Pope Leo XIII decried what he called "the unrestrained freedom of thought" that he saw running rampant as the twentieth century approached, and warned that religion and science should stay out of each other's sandboxes. In *Humani generis* (1950), Pope Pius XII said "the Church does not forbid" research and discussion related to biological evolution. But the encyclical contains a self-cancelling message typical of papal pronouncements: "Men experienced in both fields," meaning science and theology, were free to study the issue, so long as their conclusions do not contradict key religious assumptions, including the ideas that "souls are immediately created by God" and that humans cannot have ultimately come from nonliving matter.

It was not until an address to the Pontifical Academy of Sciences in 1996 that Pope John Paul II improved on Pius XII. "Today," he said, "more than a half-century after the appearance of [Pius XII's] encyclical, some new findings lead us toward the recognition of evolution as more than an hypothesis." Even so tepid a nod to science after 137 years is more than many churches have done to accept the scientific consensus on this issue.

If religion is so doggedly resistant to change, what is the use of addressing religion at all? The answer is in the title of this section. Most of the change in religious attitudes and beliefs has come not through institutions but through individuals. Just as cultural evolution sprints while genetic evolution crawls, so the attitudes and beliefs of religious individuals can sprint to the finish line before church leaders and scripture even get their shoes tied. Few religious people still support the idea of slavery, for example, even though the Bible endorses the practice and even spells out details for how to enslave others.

But slavery is a value question. What about questions of fact, like evolution? Even as many doctrinal statements continue to define clear denominational positions in favor of the special creation of humankind and against evolutionary science, individuals have moved much more quickly to accept scientific answers. Even among evangelicals in the United States, most of whom belong to churches with strong anti-evolution doctrines, fully 35 percent accept the idea that humans and other living things have evolved over time. For mainline Protestants, the number is 66 percent.[36]

So instead of relying on churches to change their doctrines, we need to continue encouraging a process that is already underway—urging individual members of religious congregations to adopt the scientific consensus on all facts as an overriding amendment to the teachings of their religion. They can continue to accept their religion's teachings on values that are not based on false "facts," and continue to enjoy the benefits of community and identity.

If most of the members of a religious congregation accept the scientific consensus on all facts, their leaders will soon follow out of self-preservation. To avoid becoming hopelessly disconnected from the perspectives of their own members, each religion will evolve to accept the conclusions of science.[37] In the coming sections, we will explore some religious organizations that have formally evolved to embrace complete acceptance of the scientific consensus on facts.

As we urge people of various perspectives and labels to fully accept the scientific consensus on the facts as an overriding amendment to any factual beliefs already present in their traditions, we should give them a simple label to refer to their method of evaluating factual beliefs. In chapter 4, we will analyze the options for this word and suggest *evidist* as the best option—not as a replacement for their current identifications, but as a modifier to their religious label that signals their overall approach to knowledge.

Using this word to self-identify, people can say, "I am an evidist Catholic," for example, or "I am an evidist Sunni Muslim," or "I am an evidist Hindu," or "I am an evidist atheist." Each person can use the adjective "evidist" to state succinctly how his or her factual views on all topics relate to science, just as they can use the adjective "secular" to state succinctly how their religious views relate to politics and public discussion. Someone who identifies as Muslim but accepts the scientific consensus and believes religion and politics should remain separate can say, for example, "I am a secular, evidist Muslim."

Though many atheists see the evidist acceptance of all conclusions of science as a way of knowing embraced solely by atheists, this is false in two respects. First is the fact that many religious people accept all the conclusions of science, even when it conflicts directly with some of the doctrines of their churches. University of Wyoming sociologist Shiri Noy describes this position as one in which both science and religion are valued by an individual, sometimes creating an inconsistency and sometimes not, depending on what they value about the religion.[38] The second reason atheists cannot lay sole claim to evidism is that some atheists are not evidists. They are inconsistent in their application of evidence-based belief formation. They may, for example, accept the evidence against existence of any gods while ignoring evidence against unscientific healthcare practices such as homeopathy or certain conspiracy theories.

If you have become an evidist and you are on the fence about leaving your religious congregation—too much supernatural nonsense, questionable morality, antiscience messaging, whatever the reason—pause for a moment to consider where you can do more good for your fellow humans. If you feel that the church is a bad social influence, then recognizing this while being a member of the congregation puts you in a unique position to help change it.

So here's the question: can you do that most effectively from the outside . . . or from the inside?

For many, staying in the congregation would be too difficult. Sometimes legitimate anger or a feeling of betrayal makes it impossible to walk through those doors ever again. They cannot be faulted for that, and leaving will create helpful pressure for change by those left behind. But if you are in a strong enough emotional and intellectual position to stay, consider doing so for the benefit of others. Your influence will be immeasurably greater than it could be from the outside. Share your views and concerns with other members of the congregation. Volunteer for committees and positions of influence. If you persuade others to question the church's beliefs or policies, they will thank you for opening that door for them. If you persuade many, the leaders will be compelled to take note and they will change sooner or later.

You may be saying to yourself that this is a pipe dream. Churches don't change, you say. Don't believe that for a minute. They can and they do. Most churches in the United States were grounded in biblical literalism two generations ago. Now most have dropped literalism. It didn't happen because their leaders accepted scientific findings; it happened because their congregants left the congregations or dragged them along. Biblical literalism in the United States dropped from 66 percent in 1963 to 30 percent in 2008,[39] and many churches quietly amended their doctrines to keep up. The same has happened as popular acceptance of evolution, women's rights, civil rights, and LGBT rights have outstripped the outdated views of the churches.

Humans are social animals and they will always need congregations, whether or not the congregations call themselves "religious." If you help your congregation change, you will do more good for all the members and for outsiders who learn of the change than if you merely vote with your feet.

As you share your views with a pastor or other leader, you might raise their level of sensitivity so they never talk about supernatural ideas when they are talking alone with you. If you achieve a critical mass of members who agree with you, your group might persuade the pastor or other leader that it is all right if they want to make god talk when talking privately with those who want to hear it, but that it is offensive to a large enough portion of the congregation that they should lay off it when speaking to the assembled congregation. Good values can be promoted in a sermon without relying on the hypothesized existence of a god. This will be progress that will help all humanity.

## USE THE WORD RELIGION IN WAYS THAT ALLOW CONSISTENCY WITH SCIENTIFIC CONSENSUS ON THE FACTS.

Because most traditional religions rely on facts that are not supported by evidence, the scientific way of knowing (evidism) undermines most traditional religions. However, there are religious leaders and congregation members who call themselves "religious" and yet fully accept the scientific consensus on the facts.[40] They are all agnostics and most of them also consider themselves to be atheists. Using the new word, these people are *evidists*. The scientific way of knowing (evidism) is not inherently inconsistent with religion, depending, of course, on how *religion* is defined.[41]

Philosophers and social scientists have grappled for centuries with suitable definitions of *religion* and have yet to reach consensus.[42] In any field of inquiry, effective definitions of terminology should further understanding by helping people keep distinguishable concepts separate in their minds. In their efforts to further this objective, some authors consider Confucianism and Buddhism to be *religions* and other authors do not. Below we discuss five recently developed religions that most of these authors would not consider to be *religions*. However, to help cultural evolution move in a good direction, we should broaden our definition of *religion* for marketing reasons and to promote equality, and these reasons for choosing a definition are more important than reasons of philosophical or sociological analysis.[43] When they want to express a meaning that is limited to religions with supernatural elements, the philosophers and sociologists can call them *nonevidist* or *supernaturalist* religions, in contrast to *evidist* religions.

In many political systems, "religions" are accorded special privileges over nonreligions. For example, in the United States, the

First Amendment to the Constitution states that "Congress shall make no law ... prohibiting the free exercise" of "religion," and various states and the federal Congress have passed statutes granting privileges to religions. These laws have been interpreted to allow churches to ignore zoning laws and to allow religious parents to keep children out of public schools and avoid vaccinations. A core principle of secularism is equality of all groups before the law. To implement this principle and treat all groups equally, either we must allow any group that wishes to claim the privileges of "religion" to do so, no matter what their beliefs or reasons for affiliation (as we argue in this book), or we can limit the religious privilege laws to cover only what people think or do that affects only themselves and has no effect on others.[44] For making legal systems more secular and treating all groups equally, it doesn't matter which is done. However, to move culture forward to accept evidism, we need to let any group call itself "religious," and this will also minimize the problem of unequal privileges for religion because any group will be able to claim the privileges.[45] If enough new groups claim religious privileges under the law, it will take away political support for the religious privilege laws.

We should change the common definition of *religion* to make it easier for most people to accept fully the facts established by science (become evidists) while still calling themselves religious. Specifically, we should broaden the definition of *religion* to encompass any group that wishes to call itself "religious," no matter what is their view of facts or what values they encourage. That is, any group with any kind of mission or purpose or philosophy could call itself "religious" and we would not say they are wrong to do so.[46] If we do this, it will give space for religions to evolve to change their positions on facts to be consistent with science (become evidists). This will make it easier for religions to get out of the business of opining on facts and limit themselves to opining on values. Several religions have already done this and they still call themselves religious. For sociologists or philosophers or politicians to contradict them and tell them that they are not "religious" will hold back cultural evolution.

People can accept the scientific way of knowing the facts and become evidists as an overriding amendment to their preferred religious or values affiliation whether it is humanist, Unitarian Universalist, Buddhist, Jewish, Christian, Muslim, Hindu, Sikh, Jain, Confucian, vegan, environmentalist, new age, pacifist, socialist, or other. When they do this, what is left of their religion is the values that are not based on false "facts." To make it easier for everyone to do this, we should not tell them they are wrong if they choose to call themselves "religious" even though they fully accept the scientific consensus on the facts (evidism).

Many New Atheist authors bash all religions. They shouldn't. They should only bash religions that are inconsistent with the scientific consensus on the facts. It is all about how we choose to use and define the word *religion*. Adopting the usage of *religion* that we urge, which some New Atheist authors already follow, will help move humanity toward the scientific way of knowing (evidism) and secularism.

Most New Atheists want everyone to abandon the religion of their heritage. This is both unachievable and unnecessary. Although large numbers may leave the traditional religions, for those who do not, our only option is to lead each religion to change to accept the conclusions of science including atheism and agnosticism.

To maximize the chances of persuading religionists to accept the scientific consensus on the facts (become evidists), secular activists should change how they use the words *religion* and *religious* so that all religions are not deemed incompatible with the scientific consensus. Some philosophers prefer to apply the word *religion* only to worldviews where there is a supernatural element because this gives the word *religion* a most useful meaning for their purposes. However, marketing of the scientific way of knowing facts (evidism) is more important than facilitating esoteric philosophical discussions. The philosophers too should change how they use the word *religion* to be consistent with how secular activists should use the word and refer to *supernaturalist* or *nonevidist* religions when that is their intended meaning.

An important path to help people accept the scientific consensus on the facts is to allow them to keep their "religious" congregations and organizations, continue to call themselves "religious," and still fully accept the scientific consensus on the facts (become evidists). Many people have done this. We should make this path easier to follow by exercising care in how we use the word "religion." We should give people as many paths to the scientific way of knowing and secularism as possible. We should not discourage people from continuing to call themselves "religious" after they have come to accept the scientific consensus on the facts.

### GROUPS WHO ACCEPT THE SCIENTIFIC CONSENSUS AND STILL CALL THEMSELVES "RELIGIOUS"[47]

*Religious Humanism*
People who fully accept the scientific consensus on the facts can have widely divergent values. Humanists accept the scientific way of knowing for determining facts. They hold particular values, rejecting other values.

Humanists have historically divided into "secular humanists" and "religious humanists," although this may be a distinction without a significant difference. According to William R. Murry, a leading author on Religious Humanism,[48] a major difference between these two groups is that the secular humanists dislike the loaded label "religious" while the religious humanists prefer to use the word.

Murry asserts that all humanists, even the religious humanists, fully accept the scientific consensus on the facts, saying: "Religious Humanists agree with the "New Atheists" in their critiques of supernaturalism and *traditional* religion. . . Religious Humanism . . . is about the *values* we stand for: human worth and dignity, human well-being, human flourishing, social justice, and equity for all people" (emphasis added).[49]

If humanists who fully accept the scientific consensus on the facts want to call themselves "religious," it would work against the objectives of the secularists and New Atheists to reject their choice of labels.

The U.S. Federal Bureau of Prisons gives special privileges to prison inmates who claim to be religious. In 2010, an inmate who claimed that religious humanism was his religion was denied those privileges in a U.S. prison. The prison officials asserted that humanism is not a religion. The inmate and the American Humanist Association sued in 2014 to obtain equal rights for humanists relative to other religions.[50] The federal judge ruled that humanism is a "religion" for purposes of prison rules.[51] In July 2015, the U.S. Bureau of Prisons accepted humanism as a "religion" and granted equal rights across the entire federal prison system.[52] This is as it should be.

Of course, it would be even better if people who claim to be "religious" had no superior rights in the prison system or any other branch of government over people who do not claim to be "religious." Allowing any group who wishes to call itself "religious" to do so will move us in this direction.

*Unitarian Universalism*
Unitarian Universalism (UUism) is an excellent example of how a Christian religion can evolve to become consistent with the scientific consensus on facts. UUism has gotten out of the business of telling people what to believe in the realm of facts and only offers guidance in values. There is no dogma in the realm of facts, no set of claims that members must believe in order to belong. Leaders and members can be theistic or not as they choose. The shared "beliefs" are all values.[53]

Although the trend in Unitarian Universalism is toward greater nontheism, only about half of Unitarian Universalist members are now nontheistic.[54] The other half still sometimes refer to "God" or to humans having a "soul." However, a growing number of congregations have a policy that any minister hired by the congregation must agree not to use "the G word" in any message to the congregation, and some congregations extend this policy to discourage references to anything supernatural, including a "soul." For advancement of acceptance of the scientific consensus on the facts, this is laudable evolution of religion.

Despite having none of the factual belief requirements of a traditional religion, most Unitarian Universalists still refer to themselves, their gatherings, and their organization as "religious." Neither activists nor governments nor anyone else should have standing to tell them they are not religious. In May 2004, the Texas Comptroller tried to do exactly that, ruling that Unitarian Universalism is not a religion under Texas law because it "does not have one system

ONE SYMBOL OF RELIGIOUS HUMANISM COMBINES THE UU CHALICE LOGO WITH THE "HAPPY HUMAN," AN INTERNATIONAL SYMBOL OF HUMANISM.

of belief" and stripping a Unitarian Universalist Church of its tax-exempt status. This decision was quickly reconsidered and reversed.[55]

Like Unitarian Universalists, other Christian denominations have made steps toward acceptance of the scientific consensus on the facts. The Catholic Church, for example, abandoned its former strict dogmas that the sun revolves around the earth, that the earth is less than a million years old, and (as previously noted) that humans did not evolve from other animal forms. This is progress—often agonizingly slow, but progress nonetheless.

*Pragmatic Buddhism*

Many Asian religions, including Taoism, Confucianism, and Buddhism, were founded as nontheistic life philosophies, only to accumulate supernatural trappings over time. Gautama Buddha specifically warned that supernatural beliefs, including the idea of gods, can create a serious obstacle to achieving *nirvana,* the total freedom from suffering. This so impressed his followers that within a few generations Buddha was venerated as a god. Taoist founder

Lao Tzu, who also spurned the idea of gods and warned against superstition, suffered much the same fate. Though many schools of Buddhism around the world continue to hold supernatural beliefs, others are entirely nontheistic, including Pragmatic Buddhism.

The Pragmatic Buddhists make an explicit claim to accept the scientific consensus fully, rejecting reincarnation and other supernatural factual assertions.[56] But are they religious? The U.S.-based Center for Pragmatic Buddhism takes, unsurprisingly, a pragmatic view of whether they call themselves "religious," depending on the meaning of that word to their audience. A former student of the Center put it this way: "I rarely speak of religion in anything other than general terms, and only as a word useful for establishing relationships or dialogue between various interfaith groups and Buddhism. . . . 'Religion' is a useful word for now as Buddhism works to achieve cultural authority. But is not necessary."[57]

*Humanistic Judaism*

Humanistic Judaism is a nontheistic religion that "combines attachment to Jewish identity and culture with a human centered approach to life."[58] Its leaders espouse no positions on factual issues that are inconsistent with the scientific consensus, and most or all members are nontheistic. With more than 40,000 members in the United States alone, Humanistic Judaism is not a heresy but one of the five officially recognized branches of Judaism today.

Although the leaders of Humanistic Judaism fully accept the scientific consensus on the facts, they also claim that Humanistic Judaism is a religion because it is "a set of beliefs [about *values*] to which people hold fast." Rabbi Sherwin Wine, the founder of Humanistic Judaism, said it is an "ancestral religion" rather than a "salvation religion." "Humanistic Judaism is also a religion in its structure: its congregational model, school for children, adult education, and life cycle ceremonies all follow the religious model."[59]

*Ethical Culture*

Ethical Culture is an association of nontheistic congregations called Ethical Societies organized under the name American Ethical Union.

THE "HUMANORAH" OF HUMANISTIC JUDAISM COMBINES THE JEWISH MENORAH AND THE HAPPY HUMAN OF HUMANISM. COPYRIGHT SOCIETY FOR HUMANISTIC JUDAISM (WWW.SHJ.ORG). USED WITH PERMISSION.

They promote no factual beliefs that are inconsistent with the scientific consensus. The organization and congregations call themselves religious in the same sense as Humanistic Jews and UUs, a meaning reflected in statements like these: "Ethical Culture believes that a sense of the religiosity emerges from more sensitively recognizing, appreciating, evoking and celebrating the humanity that resides in all people."[60] "Religion is that set of [value] beliefs and/or institutions, behaviors and emotions which bind human beings to something beyond their individual selves and foster in its adherents a sense of humility and gratitude that, in turn, sets the tone of one's world-view and requires certain behavioral dispositions relative to that which transcends personal interests."[61] Ethical Culture is treated as a religion for government taxing and regulatory purposes.

*Liberal Quakerism*
The Liberal Quakers have no shared creed, no central scripture, and no dogma. Like UUs, some are theistic and some are not, but even those who believe in a god share one important value: no person can

tell any other person what the experience of God is like.[62] When you dispense with creed and dogma, there is no impediment to the full embrace of the consensus on science, and most Liberal Quakers do just that.

A nonreligious person who categorically rejects the possibility of finding common cause with the religious on issues related to science allows a largely semantic distinction to thwart the building of powerful and effective alliances among people who share a vital opinion—that the scientific consensus is worth adhering to.

### THEOLOGIANS WHO URGE THAT CHRISTIANITY EVOLVE TO FULLY ACCEPT THE SCIENTIFIC CONSENSUS

Many Christian theologians have come to accept the scientific consensus on the facts.

In 1967, a Christian educator and Presbyterian minister, Sir Lloyd George Geering, published his views that there is no God and there was no resurrection, for which he was charged with heresy. From 1968 to 2014, Geering published over a dozen books explaining how the important elements of Christianity do not require any supernatural beliefs and urging that the religion be updated to be consistent with all scientific knowledge.[63]

In 2014, Dr. Daniel C Maguire, a Catholic educator and theologian with a degree in Sacred Theology from one of the world's major Catholic universities, the Pontifical Gregorian University in Rome, published *Christianity without God*. Dr. Maguire argues that traditional, supernatural aspects of Christianity can be comforting but have been undermined by science, have questionable roots in historical traditions, and are unnecessary. His view is that the moral epic of the Hebrew and Christian Bible nevertheless offers a useful ethic for all time.[64]

In *The Gospel of Christian Atheism* (1996), theologian Thomas J. J. Altizer urged people to focus on the moral message of Jesus while fully embracing the factual message of modern science.

Then there's Bishop John Shelby Spong,[65] probably the most radical prominent clergyman of all time, who urges Christianity to "change or die." The changes he suggests include recognizing that no supernatural God exists, that all miracles are false, and that scripture bears limited ethical relevance in modern life.

Gretta Vosper of Toronto is the most prominent and articulate of many pastors with active congregations that are leading all who listen to make modern religion fully consistent with evidism. Her book, *With or Without God: Why the Way We Live Is More Important Than What We Believe*, published in 2008, explains how Christianity can be updated to be fully evidistic.[66]

### RELIGIOUS SCIENTISTS WHO URGE RELIGIONS TO EMBRACE ALL OF SCIENCE

Professor Tom McLeish, a materials scientist at Durham University in England, undertook an extensive theological interpretation of the Judeo-Christian tradition.[67] While he presently believes that the "cosmos is shot through with meaning"[68] provided by a creator and humans are made in the image of God,[69] he argues that a correct reading of the Bible, especially the Old Testament, urges an unlimited scientific exploration of the natural world, that theology must evolve to accommodate what is discovered, and that theology must encompass all science. While his position at the time of his writing did not yet go all the way to *evidism* because of his religious beliefs not supported by evidence, it is an important step in the right direction and would make a big difference if all religions were to follow his urging.

### EXAMPLES OF "NEW ATHEIST" AUTHORS BEING TOO NARROW IN THEIR USE OF THE WORD "RELIGION"

In the examples of New Atheist writing that follow, we have no disagreement with the substance stated by the author. But their generalized use of the word "religion" unjustly implicates many people and religious expressions, including those listed above, in their critiques. If they were to consistently insert a qualifier such as "most" or "traditional" in front of the word "religions," they would give important support to religious humanists and other groups who fully accept both secularism and the scientific consensus on the facts and yet prefer to call themselves "religious." Such a practice also enhances their critique by accurately recognizing that exceptions exist. This would make it easier for millions of others to become evidists while still referring to themselves as "religious." In each quotation below we have modified the text with brackets and italics to make clear the simple but important change we suggest.

*Richard Dawkins, author of* The God Delusion

> Certainly I see the scientific view of the world as incompatible with [*nearly all*] religion.[70]

*Sam Harris, author of* The End of Faith

> [*Traditional*] religion unites, motivates, and consoles beleaguered people not with knowledge, but with superstition and false promises.[71]

> There is an epidemic of scientific ignorance in the United States. . . . But it would seem that few things make thinking like a scientist more difficult than [*traditional*] religion.

> [*Most of*] the world's religions are predicated on the truth of specific doctrines that have been growing less plausible by the day.[72]

> . . . my rather incessant criticism of [*traditional*] religion in my books, articles, and lectures.[73]

*Christopher Hitchens, author of* God Is Not Great

> People are frightened of death, and the central lie of [*nearly*] all religion is that there's a cure for this and an exception we've made in your own case: an eternal life offered if you make the right propitiations and the right abjections.[74]

> I am not even an atheist so much as an antitheist; I not only maintain that [*nearly*] all religions are versions of the same untruth, but I hold that the influence of churches and the effect of [*most*] religious belief, is positively harmful.[75]

*Daniel Dennett, author of* Breaking the Spell

> I think that there are no forces on this planet more dangerous to us all than the fanaticisms of fundamentalism, of all the species: Protestantism, Catholicism, Judaism, Islam, Hinduism, and Buddhism, as well as countless smaller infections. Is there a conflict between science and [*nearly all*] religion here? There most certainly is.[76]

*Alex Rosenberg, author of* The Atheist's Guide to Reality

> Just think about religion, any religion. [*Nearly*] every one of them is chock full of false beliefs.[77]

Some atheists suggest that qualifying language of this kind is offered out of a desire to placate religious believers. Not true; it is offered in the interest of accuracy. When Alex Rosenberg asks us to think about "any religion," and we think about Unitarian Universalism, or Ethical Culture, or Liberal Quakerism—none of which is organized around factual beliefs at all—his statement is instantly falsified. If truth is our goal, we should make a careful practice of avoiding false statements.

### EXAMPLE OF A NEW ATHEIST AUTHOR WHO LETS "RELIGION" BE CONSISTENT WITH THE SCIENTIFIC CONSENSUS ON THE FACTS

*Jerry A. Coyne, author of* Why Evolution Is True *and* Faith Versus Fact: Why Science and Religion Are Incompatible

> Now I am not claiming that all faith is incompatible with science and secular reason—only those faiths whose claims about the nature of the universe flatly contradict scientific observations. Pantheism and some forms of Buddhism seem to pass the test. But the vast majority of the faithful—those 90 percent of Americans who believe in a personal God, most Muslims, Jews, and Hindus, and adherents to hundreds of other faiths—fall into the "incompatible" category.[78]

The simple addition of "most" to Coyne's last sentence made all the difference.

## SHOW THAT YOU ARE WILLING TO STATE PUBLICLY YOUR ACCEPTANCE OF SCIENCE AND THE IMPORTANCE OF RELIGIONS CHANGING TO BE CONSISTENT WITH THE FACTS.

Among experts in persuasion, it is accepted wisdom that, when speaking about facts, the credibility of a speaker increases as he or she moves beyond a small audience to a larger public hearing. One reason is that people assume, consciously or unconsciously, that a large audience includes a number of potential whistleblowers who will speak up and challenge anything that is wrong, so the speaker will be more careful than when speaking to only a few.[79]

This doesn't apply to value statements. As noted above, when speaking about values, you should always be ready for credible disagreement. The greater benefit of the large, public audience only applies to factual statements. You can assert with confidence any of the facts established by consensus of science, including the facts listed in the second section of chapter 3.

If you want to convince your friends or family of something, they will naturally be more likely to find your assertions persuasive if you make them publicly or to a large audience instead of merely sharing them with a small group in your living room. The same may apply to statements made in your real name on publicly accessible websites. By leaving the cocoon of those who have always found you smart and charming, you decrease the chance of audience-size prejudice diminishing your credibility.

## EXPLAIN THE CORRECT MEANINGS OF THE WORDS AGNOSTIC AND ATHEIST.

The English biologist Thomas Henry Huxley probably didn't mean to introduce confusion into the world of belief labels when he coined the term *agnostic* in 1869. He was simply solving a problem for himself—

the problem of accurately describing his own beliefs.

The word *atheist* seemed to Huxley to imply certainty on a question no one can ever be certain about. So, although he did not believe, he wanted a way of underlining his lack of total certainty.

After puzzling over the problem for some time, Huxley came up with the word *agnostic*—Greek for "not knowing"—to describe his position. The term quickly became popular, and just as quickly misunderstood.

Most agnostics are assumed to be in the shrugging middle, flipping a coin on the question of God, leaning neither one way nor the other. In fact, very few agnostics would describe themselves in those terms. Most, in fact, have beliefs that are indistinguishable from the average atheist. They simply choose to emphasize their doubt, not their confidence.

To draw out this nuance, Richard Dawkins—who, like Huxley, is also an English biologist—created a seven-point belief scale between the extremes of certainty.[80] (See figure 6.) A "1" on the scale goes to the person who claims with absolute certainty that God exists. No new information can ever change that opinion. On the other end of the scale, at "7," is the person who claims to be certain that God does *not* exist and is likewise immune even to the possibility of a changed mind.

Someone who scores a "2" believes God is very probable, but stops short of claiming absolute certainty, while "3" indicates a less confident belief in God but more likely than not.

On the other side, a "6" indicates a strong probability that God does *not* exist, but likewise stops short of absolute certainty. This is the position supported by the evidence and the scientific way of knowing, evidism. A "5" indicates a greater probability of a god, while still believing the chances are less than 50 percent.

This brings us to the number "4." Here, precisely in the middle, is what most people think of as the agnostic position at 50–50. But you'd be hard-pressed to find an agnostic who genuinely sits on that fence, just as it is the rare atheist who expresses total certainty at position 7.

In a 2012 interview with the Archbishop of Canterbury, Dawkins was asked why he doesn't call himself an agnostic if he says he isn't completely sure God doesn't exist. When he said that he does consider

## FIGURE 6. THE DAWKINS SCALE OF THEISTIC PROBABILITY

### 1. STRONG THEIST
*I do not believe, I know God exists.*

### 2. DE FACTO THEIST
*I don't know for certain, but I strongly believe in God.*

### 3. LEANING TOWARD THEISM
*I am very uncertain but am inclined to believe in God.*

### 4. COMPLETELY IMPARTIAL
*God's existence and nonexistence are equally probable.*

### 5. LEANING TOWARD ATHEISM
*I do not know whether God exists but I'm skeptical.*

### 6. DE FACTO ATHEIST
*I dont know for certain, but I think God is very improbable.*

### 7. STRONG ATHEIST
*I know there is no God.*

himself an agnostic, the audience gasped. The world's most famous atheist had admitted he is actually an agnostic!

Of course, saying he "admitted" to being an agnostic is as inaccurate as saying a Christian "admitted" he is actually a Methodist. In both cases, two labels emphasize different things. Dawkins is confident that no gods exist (he is an atheist), but he also knows that no one can be sure of such a thing (he is also an agnostic).

Agnosticism is one of the most powerful and important ideas in human knowledge. Making its meaning more widely known, and how it relates to the word *atheist,* will help us spread both science and secularism to our friends and the world.

Every fact has a probability of being true that is less than 100 percent. Because nothing is known with 100 percent certainty, all who fully accept the scientific way of knowing facts (evidists) can be accurately labeled agnostics, meaning they recognize that all knowledge is uncertain to varying degrees. If they always act for all purposes as if no god exists, they are also atheists, which simply means "don't believe in gods." All atheists who accept the scientific way of knowing (evidists) are also agnostics because they do not know with 100 percent certainty that no gods exist.

An atheist doesn't have to claim certainty. He or she simply says "the chances that a god exists are so small that I act for all purposes as if there is no god." This misunderstanding is spread by leaders of religions who, out of self-interest, try to persuade people that if you are not 100 percent sure there are no gods, then you are an agnostic, not an atheist. This is wrong: most people who are one are also the other.[81] Atheists who think they know with 100 percent certainty that no gods exist, and therefore decline to call themselves agnostics as well, are not following the scientific way of knowing and are therefore not evidists. All evidists are agnostics and, though they may not agree to the label, because they never determine their actions based on the possibility that there might be a god, they are also atheists.

## REPLACE MOST USES OF THE WORD ATHEISM WITH ANTITHEISM, OR HUMANISM, OR EVIDISM.

Many people casually use the word *atheism* with an implication that it is a philosophy and that people who are atheists share important beliefs or values. But *atheist* only means "doesn't believe in any gods" and, more concretely, "never bases any decision or action on the possibility that there might be a god."[82] Though, like any group that shares a significant feature—teachers, daredevils, athletes, Republicans—there are some common trends and tendencies within the group, atheists hold a divergent range of values and even beliefs about facts. They no more share a philosophy with one another than people who don't believe in leprechauns share a philosophy. Atheism has almost zero content and, because of this, should rarely be referred to as an "ism" at all.

Some people use the word *atheism* to refer to a social activism movement that would like to eliminate religion from the world. This movement, commonly called *New Atheism,* has a lot of substantive content and certainly qualifies as an "ism." However, the most accurate word for it is *antitheism,* a term favored by one of the movement's most eloquent spokespersons, the late Christopher Hitchens. In an article in *Salon,*[83] author and religious scholar Reza Aslan traces the long history of this movement and argues persuasively that the label *New Atheism* is a misnomer because the word *atheism* is empty, having almost no meaning, and the movement is not new. He correctly argues that the best word is *antitheism.*

Some people use the word *atheism* in a way that assumes all atheists accept all science and hold humanistic values, neither of which can be assumed. Where that is the intended meaning, they should use the word *humanism* instead. Using the word *atheism* fails to get their meaning across and perpetuates misunderstanding of the meaning of this word.

Some people use the word *atheism* to refer to a philosophy that says all we can know is based on evidence from the natural world— that nothing supernatural exists. This philosophy says nothing about values. A clearly defined word already exists for this philosophy. It is called *naturalism.* Unfortunately, as we explain in detail in chapter 4, if you try to teach your friends and family this meaning of the word *naturalism* and urge them to identify as naturalists, they will immediately picture David Attenborough or Jane Goodall, pith-helmeted in the jungles of Africa. To them, a naturalist is an expert on the plants, animals, and ecosystems of the natural world. This usage is unfortunately so deeply ingrained that a new word is needed to spread the philosophy of naturalism to friends and family with less confusion. That new word is *evidism.* You should urge your friends and family to identify as evidists—people who embrace the scientific way of understanding the world. And, if they retain values from a religion, they can continue to identify as "religious" at the same time.

## BE ALERT TO OTHER MEANINGS
## OF THE WORD SECULAR.

In chapter 1, we discussed the core values meaning of the word *secular*. Everyone agrees with the core values meaning that, for people to be *secular*, they must avoid relying on supernatural religious factual assertions in public policy discussions. All the arguments they make about what government should do or not do must be based on facts that people of all religions or no religion can agree on. Falling within this core values definition, people can be secular Christians, or secular Muslims, or secular Jews, or secular Hindus, and so on.

But there are authors and activists who use the word *secular* in a way that requires more than this for a person to be considered *secular*. They do not agree with one another on what else you must not do or not believe to be considered *secular*. Some of these authors say there is no such thing as a secular Christian or secular Muslim or secular Jew. Their usage is synonymous with "nonreligious." These authors variously require one or more of the following:

- not be a member of a religious congregation;
- say to others that you are "not religious";
- not display religious symbols;
- not believe in your heart of hearts in any gods or demons;
- not pray, and not pretend to pray when others are watching;
- say to yourself that you are a *freethinker* or an *agnostic* or a *nontheist* or an *atheist*.

As we explained above, it is a reality of politics—of people getting along with each other—that everyone should be *secular* within the core values meaning of the word. For sharing reality with friends and family and the world, this is essential. It is an evidistic statement of fact that, unless we all adopt the values of secularism, we will never have a peaceful and happy society.

As for the other six elements that some authors say everyone should adopt to be *secular*, we quite disagree with the first three.

Although our disagreement with these authors is merely semantic, semantics can make a big difference in marketing.

First, many people find great satisfaction in congregational belonging. Rather than leaving their congregations, as we explained above, we urge people to change the congregations from within to become secular, within the core values meaning, and to accept the consensus of science on all facts (evidism).

Second, as people change their congregations from within, it is almost always impractical and unachievable to drop the word *religious*. Instead, as we explained above, we should help the meaning of the word *religious* evolve so that a person can be *religious* and *secular* and *evidist* at the same time.

Third, while religions are evolving to accept the core of secularism (as many have already) and the conclusions of science on all facts (as some have already), there is no reason for people to abandon religious symbols with which they identify, provided the symbols are not inherently inconsistent with secularism or the conclusions of science (evidism).

As for the last three elements listed above, we think you should try to persuade your friends and family to adopt them because, as we explained above, they are required if you accept the conclusions of science.

## Point out subtle supernatural and tribalistic messages in mass media.

Evolution has made humans into supremely social animals with high social intelligence. Consequently, we have nearly limitless interest in gossip and stories about people and about fictional anthropomorphic creatures. This is a good thing in one major way: cultural evolution makes progress in large part through popular stories that people consume in movies, books, and other mass media.

Because movies, books, and what celebrities say are major influencers of cultural evolution, we should help others recognize and think about the subtle messages that come from these sources to help advance secularism and acceptance of the scientific consensus on the facts (evidism). Because of our personal capital with those who know us well, our comments can help friends and family learn to spot these subtle messages. And our public comments, in online forums and elsewhere, can raise the awareness of authors, bloggers, politicians, and other thought leaders regarding subtle messages in their own words that they may not have been aware of.

The philosopher and humanist Corliss Lamont described just such a consciousness-raising process in his own career. After decades using the words *man* and *mankind* to describe all humanity, and *he* as the default for a generic person, Lamont was approached by his wife and a female assistant who asked if he would consider switching to gender-neutral language instead. When he rebuffed the suggestion, the women picked up a volume of his work and read it aloud, changing specifically male references to specifically female ones. So the passage

> If you asked an American of the nineteenth century what the future of mankind might hold, he might have described a global empire, with nature subdued and man triumphant.

becomes

> If you asked an American of the nineteenth century what the future of womankind might hold, she might have described a global empire, with nature subdued and woman triumphant.

Lamont described the shock of realization that swept over him, and his work made use of gender-neutral language for the remainder of his career.[84]

It is just this kind of "shock of realization" that can help us raise the consciousness both of family and friends and of thought leaders regarding subtle supernatural and tribalistic messages.

For movies or books or other mass media in which a supernatural or tribalistic message is obvious—science fiction that contradicts science, for example, or overtly religious stories—we need not speak out. Everyone is aware of such overt messages, whether or not they agree with them. It is the subtle messages that often need to be pointed out for that "shock of realization" to occur. The unwanted conflict that results from supernatural explanations and tribalistic thinking should be enough to motivate us.

Trying to stop the creation of such stories would be both hopeless and contrary to the spirit of free expression. As long as there is a demand, the supply will continue, and that's fine. A lot of great literature and popular entertainment includes these subtle messages. The *Star Wars* films promote belief in the supernatural by invoking "The Force." The *Lord of the Rings* trilogy promotes tribalistic thinking and emotions with an us-versus-them mentality in the characterization of Mordor and its citizens. But rather than try to put these or future genies back in their bottles, our efforts are best spent encouraging people to be awake to the supernatural and tribalistic messages they carry.

## ENCOURAGE EDUCATION.

No person can assemble on their own enough valid knowledge to adopt the scientific way of knowing without education. We all must learn from others who shared their contributions through writing and correcting one another's contributions to reach a scientific consensus. By soaking up knowledge from appropriate sources, individuals can move their thinking toward the scientific way of knowing (evidism) without being explicitly aware of the difference between the scientific way of knowing and the intuitive ways of knowing that they are born with.

Only through education consistent with reason and good evidence can people get beyond their natural tendencies to deceive themselves and to accept uncritically whatever their elders tell them. Preferably, the education includes epistemology—the study of sources of knowledge and how we can know what to believe. Learning science and how to interpret evidence is entirely rational, but that doesn't mean it comes naturally. It requires study.[85]

Something rather profound happens when people come to understand and accept the scientific way of knowing (evidism): they cross a bright line and are extremely unlikely to slip back into a solely natural or intuitive way of knowing. But this education, once achieved, isn't passed on genetically; it has to be achieved again and again with each generation. If the level of education falls in a culture, young people can and will hold on to the intuitive ways of knowing they are born with, and the culture as a whole can easily slip back.

The best education doesn't simply fill a person's head with "the answers"; it also builds an individual's skill in separating the wheat from the chaff on their own. By showing that the most current, generally accepted scientific theories explain almost everything that matters to each person, and showing that theories that lack scientific consensus (such as nonevidist religion, astrology, and homeopathy) have so little chance of being right that they should be summarily dismissed, each person can be given the mental tools to adopt the scientific way of knowing (evidism) and to use those tools to continue making discernments long after formal education is done. With these tools, people can avoid reliance on unworthy, self-invented theories of truth that spring from their intuition or are presented to them from unreliable sources.

It is not just the common subjects of science that should be taught in schools, but also the scientific consensus on whatever each person finds interesting—what they care about in their daily lives, including religions, spirituality, and superstition. Unfortunately, teaching the science on these subjects is taboo in most schools.

## Notes

1. Daniel Dennett, *Breaking the Spell: Religion as a Natural Phenomenon* (New York: Viking, 2006). Hank Davis, *Caveman Logic: The Persistence of Primitive Thinking in a Modern Word* (Amherst, NY: Prometheus Books, 2009).

2. Daniel Dennett, *Breaking the Spell,* explains this process of cultural evolution. Rebecca Goldstein and Andy Norman argue that evolution of the brain also gave all humans a "mattering instinct" that instinctively causes them to feel that their own lives matter and gives them a strong desire that their lives matter to other humans and, cosmically, to gods or the universe, and that this provides strong motivation to believe in gods and everlasting

life. Goldstein, "Mattering Matters," and Newman, "The Mattering Instinct," *Free Inquiry* 37, No. 2 (February/March 2017): 14–27 (https://www. secularhumanism.org/index.php/3532).

3.   Here we are using the word "homeopathic" in the sense intended by the coiners of the word and advocates of the theory, specifically including the ideas that (1) more dilute remedies are more effective than less dilute remedies, and (2) the water used to dilute the remedy has some kind of memory to carry the benefits of the active component as it is diluted. In the United States, there are remedies that are labeled "homeopathic" even though they do not fit this definition. and some of these remedies are effective by more than placebo effect. The reason this word is used on these products in the United States is to avoid the requirements of the U.S. Food Drugs and Cosmetics Act which require that products labeled to treat diseases be proven safe and effective (to a very high certainty but less than 100 percent) unless they are also labeled as "homeopathic." When the FDC Act was passed, this exception was included for remedies labeled "homeopathic" as a result of antiscientific lobbying.

4.   Richard Dawkins, *Unweaving the Rainbow* (Boston: Houghton Mifflin, 2000), 142–43.

5.   Satoshi Kanazawa, "Why Atheists Are More Intelligent Than the Religious," *Psychology Today* (April 11, 2010) http://www.psychologytoday. com/blog/the-scientific-fundamentalist/201004/why-atheists-are-more-intelligent-the-religious.

6.   In *Denying to the Grave: Why We Ignore the Facts That Will Save Us* (Oxford University Press 2017), Sara Gorman and her father Jack Gorman argue that "the most valuable thing students can learn in science class is how to evaluate whether or not a claim is based on evidence that has been collected by careful use of the scientific method", with emphasis on "process over content: how science is conducted, how scientists reach conclusions, and why some things are just chance and others are reproducible facts of nature." (Kindle Locations 3462-3476). We agree this is important, especially for students. However, in our view, what is more important, for students and especially adults, is to learn that nothing is known for certain, that, with respect to facts of reality "true" and "false" are shorthand for "very likely true" and "very likely false," and that the facts claimed by scientific consensus are reliably true so we can and should all accept them.

7.   Sometimes the scientific consensus is based on robust evidence and sometimes it is only believed to be a 51 percent probability of being right. In all cases it is worthy of reliance because, until it is updated, it is the best we've got. But if you are considering making an investment in reliance on the

present best evidence, you might want to wait until the evidence is robust.

8. Wikipedia is frequently maligned as a source because its content is open to continual change. But continual change also means continual correction, and a 2005 study published in *Nature* found that such a process leads to a level of accuracy nearly as high as *Encyclopaedia Britannica*. Several subsequent studies looked at subsets of articles related to selected professional fields and found similar high rates of accuracy. See the original Nature article at http://www.nature.com/nature/journal/v438/n7070/full/438900a.html. Accessed March 17, 2015.

9. https://simple.m.wikipedia.org.

10. See Coel Hellier, http://coelsblog.wordpress.com/2014/09/30/musings-on-gettier-and-the-definition-of-knowledge-2/. The exception is mathematical "facts," so we are only talking here about empirical facts.

In an interview with the BBC, Richard Feynman stated: "I can live with doubt, and uncertainty, and not knowing. I think it's much more interesting to live not knowing than to have answers which might be wrong. I have approximate answers and possible beliefs and different degrees of certainty about different things. But I'm not absolutely sure of anything." Jerry A. Coyne, *Faith versus Fact: Why Science and Religion Are Incompatible* (New York: Viking, 2015), 38.

11. Except for purposes of scientific or philosophic inquiry where we may, during the investigation, take quite seriously for a brief time hypothesized facts with a low probability of being true.

12. See Richard Dawkins, *The God Delusion* (New York: Houghton Mifflin Harcourt, 2006).

13. This is a common belief that predates theism and eastern religions in many parts of the world, particularly Africa and Madagascar.

14. Science can provide useful data and analysis for making moral determinations and shows that many claimed bases for morality are ill founded. However, science determines no content for human values, morality, or ethics. Science shows that genetic evolution of humans as social animals gave humans a predisposition for certain values, that values evolve as culture evolves, and that there is no valid source of values outside of genetic and cultural evolution and human thought. Steven Pinker, *The Blank Slate* (2002); Joshua Greene, *Moral Tribes*.

15. David Hume is generally credited with originating fact-value separation in the eighteenth century.

In 1999, Stephen Jay Gould published a book on just this subject entitled *Rocks of Ages: Science and Religion in the Fullness of Life*. Just as we do here, Gould argued that religions should completely defer to science on all questions of fact ("the conclusions of science must be accepted a priori, and religious interpretations must be finessed and adjusted to match"). As we do here, Gould argues that science should be confined to the facts and *religion* should be confined to values, referring to each as a magisterium that does not overlap the other (NOMA, Non Overlapping MagisteriA). Gould uses the word *religion* broadly to encompass all values, including both theistic and nontheistic positions on ethics, morality, and meaning for human lives. We agree that *religion* should be given this broad a definition. Some critics interpret Gould as also arguing that science should not examine some factual questions that are important to religions, but we cannot find in Gould's book any passages that support their criticism. In his choices of examples and phrasing, Gould is, as much as possible, making nice with existing religions, apparently for practical political reasons, but when you examine each of his assertions, including the examples, he does not give religions any quarter to conflict with science or determine any facts. "Progress cannot be impeded indefinitely, and if theology does not yield its former control over the proper magisterium of science, then religion, with all its virtues, will die." Gould further asserts that science can illuminate issues of "human purposes, meanings, and values" and that the "proper place" to "seek solutions to questions of morals and meanings" is "within ourselves" rather than deities. Jerry Coyne criticizes Gould's NOMA proposal, saying, "It requires the homeopathic dilution of religion into a humanistic philosophy devoid of supernatural claims, and it gives to religion sole authority over moral and philosophical issues that have nevertheless had a long secular history" (Coyne, *Faith versus Fact*, 107.) The first critique is correct—Gould says religion must become nothing more than a humanistic philosophy or it will die—and we see this as a positive, not a negative. The second critique is not correct— Gould explicitly mentions secular development of values but then focuses on religion because it historically has been the dominant player in this arena—and this critique will become moot when religions change to the point that any difference between secular and religious analysis of values is immaterial. Coyne asserts that Gould's NOMA approach "requires both physical and metaphysical inquiry." To the contrary, Gould does not use the word "metaphysical." Gould's approach requires a values/ethical/moral inquiry, not a metaphysical inquiry. Coyne's most on-target critique of Gould's NOMA proposal is that it is overly "utopian" and wishful-thinking. Few religions will accept the changes that Gould demands in foreseeable generations. We agree that Gould did whitewash the present-day problems of religions—he seemed to be trying to create a self-fulfilling description— and he ignored the difficult problem of how to get from the present day to his proposed NOMA solution. Still, his proposed solution is the right one, and no one has yet presented an effective plan for how to get there.

16. http://www.brainyquote.com/quotes/quotes/s/stephenhaw627129. html.

17. http://www.brainyquote.com/quotes/quotes/j/johndrock147466. html?src=t_i_believe.

18. Neil Van Leeuwen proposes that, for purposes of analysis, we further subdivide beliefs about facts into two subcategories, facts about objective reality and facts about the supernatural. Beliefs in facts about the supernatural are based on teachings of different authorities from facts about reality, and they are not as much subject to falsification by evidence as facts about reality. Neil Van Leeuwen, "Religious Credence Is Not Factual Belief," *Cognition* 133 (2014) 698–715.

19. See note 3 in this chapter.

20. Thank you to Candice Wu for providing this example and offering helpful critique of the ideas in this book.

21. Coyne, *Faith versus Fact*, 236.

22. In *Why Truth Matters* (London: Continuum, 2007), Ophelia Benson and Jeremy Stangroom claim that "inquiry, curiosity, interest, investigation, explanation-seeking, are hugely important components of human happiness" and quote Richard Dawkins: "The feeling of awed wonder that science can give us is one of the highest experiences of which the human psyche is capable. It is a deep aesthetic passion to rank with the finest that music and poetry can deliver." http://www.butterfliesandwheels.org/books/why-truth-matters/extracts/.

23. For the 2013 UN World Happiness Report, see http://unsdsn. org/wp-content/uploads/2014/02/WorldHappinessReport2013_online. pdf, and for data from Pew Research Center's Global Attitudes Project, see http://www.pewglobal.org/2013/05/01/spring-2013-survey/. For additional discussion on the data and graph, see https://whyevolutionistrue.wordpress. com/2016/01/15/religiosity-is-correlated-with-unhappiness/.

24. William Masters, quoted in "We Add a Dimension to Sexuality," *Life* (June 24, 1966): 49.

25. Elizabeth Thompson Gershoff, "Corporal Punishment by Parents and Associated Child Behaviors and Experiences: A Meta-Analytic and Theoretical Review," *Psychological Bulletin* 128, No. 4 (July 2002): 539–79. Full text available online at http://www.comm.umn.edu/~akoerner/courses/4471-F12/Readings/Gershoff%20(2002).pdf.

26. Here is a statement of the values of secular humanism:

- We are committed to the application of reason and science to the understanding of the universe and to the solving of human problems.

- We deplore efforts to denigrate human intelligence, to seek to explain the world in supernatural terms, and to look outside nature for salvation.

- We believe that scientific discovery and technology can contribute to the betterment of human life.

- We believe in an open and pluralistic society and that democracy is the best guarantee of protecting human rights from authoritarian elites and repressive majorities.

- We are committed to the principle of the separation of church and state.

- We cultivate the arts of negotiation and compromise as a means of resolving differences and achieving mutual understanding.

- We are concerned with securing justice and fairness in society and with eliminating discrimination and intolerance.

- We believe in supporting the disadvantaged and the handicapped so that they will be able to help themselves.

- We attempt to transcend divisive parochial loyalties based on race, religion, gender, nationality, creed, class, sexual orientation, or ethnicity, and strive to work together for the common good of humanity.

- We want to protect and enhance the earth, to preserve it for future generations, and to avoid inflicting needless suffering on other species.

- We believe in enjoying life here and now and in developing our creative talents to their fullest.

- We believe in the cultivation of moral excellence.

- We respect the right to privacy. Mature adults should be allowed to fulfill their aspirations, to express their sexual preferences, to exercise reproductive freedom, to have access to comprehensive and informed health-care, and to die with dignity.

- We believe in the common moral decencies: altruism, integrity, honesty, truthfulness, responsibility. Humanist ethics is amenable to critical, rational guidance. There are normative standards that we discover together. Moral principles are tested by their consequences.

- We are deeply concerned with the moral education of our children. We want to nourish reason and compassion.

- We are engaged by the arts no less than by the sciences.

- We are citizens of the universe and are excited by discoveries still to be made in the cosmos.

- We are skeptical of untested claims to knowledge, and we are open to novel ideas and seek new departures in our thinking.

- We affirm humanism as a realistic alternative to theologies of despair and ideologies of violence and as a source of rich personal significance and genuine satisfaction in the service to others.

- We believe in optimism rather than pessimism, hope rather than despair, learning in the place of dogma, truth instead of ignorance, joy rather than guilt or sin, tolerance in the place of fear, love instead of hatred, compassion over selfishness, beauty instead of ugliness, and reason rather than blind faith or irrationality.

- We believe in the fullest realization of the best and noblest that we are capable of as human beings.

See http://secularhumanism.org/index.php/12. A more complete statement can be found here: http://instituteforscienceandhumanvalues.com/articles/neo-humanist-statement.htm.

27. The values are stated as follows: "We, the member congregations of the Unitarian Universalist Association, covenant to affirm and promote:

1. The inherent worth and dignity of every person;

2. Justice, equity and compassion in human relations;

3. Acceptance of one another and encouragement to spiritual growth in our congregations;

4. A free and responsible search for truth and meaning;

5. The right of conscience and the use of the democratic process within our congregations and in society at large;

6. The goal of world community with peace, liberty and justice for all;

7. Respect for the interdependent web of all existence of which we are a part."

See http://www.uua.org/beliefs/principles/.

28. As religions adapt to knowledge of human evolutionary psychology

and what can be changed through cultural evolution and they drop the assumptions that values come from a deity or from nature outside of humans, this is likely to cause cultures and religions to evolve to be more alike. Tom Clark, http://www.naturalism.org/scientism.htm#resources and http://www.naturalism.org/systematizing_naturalism.htm.

29. Bertrand Russell, "Christian Ethics," from *Marriage and Morals* (1929).

30. Justin Kruger and David Dunning, "Unskilled and Unaware of It: How Difficulties in Recognizing One's Own Incompetence Lead to Inflated Self-Assessments," *Journal of Personality and Social Psychology* 77, No. 6 (1999): 1121–34.

31. There are notable exceptions to evolution having given humans good intuitive values for interactions with families and within the tribe. Like other social animals, we evolved a violence-based social dominance behavior, as seen in all the apes, notably including our closest relatives, the chimpanzees, whereby physical force determines who leads and who gets first choice of mates or food, etc. Through cultural evolution, we have been moving away from might makes right to now discourage: sexism, including honor killings, in gender relations within families; differing values regarding natural-born versus adopted children; treating first-born sons differently; homophobia (abusing/disowning gay children by parents and siblings); and the intuited notion that children should be disciplined by beating them or starving them. We are also moving away from the naturally evolved tribal practice of creating hierarchies between family units that result in unjust social stratification of power, obedience, wealth, and rights.

32. David Lahti, "Why Does Religion Keep Telling Us We're Bad?" *Guardian* (UK) (November 22, 2011).

33. Direct quotation from Berlinerblau, *How to Be Secular,* 192.

34. Edward O. Wilson, *The Meaning of Human Existence.* In *The Necessity of Secularism,* 177, Ronald A. Lindsay identifies seven roles that religion plays in people's lives: (1) source of community, (2) source of courage, (3) means of being in harmony with the order of the universe, (4) point of contact with the transcendent, (5) meaning for their lives, (6) sense of peace and tranquility, and (7) inspiration to act morally.

35. This position is argued in detail by Stephen Jay Gould in *Rocks of Ages.* Andy Norman argues that humans have an instinctive desire to feel that their group is special, superior to other groups (part of the "mattering instinct"), and that religions cater to this instinct. Andy Norman, "The Mattering Instinct" (https://www.secularhumanism.org/index.php/articles/

8608). This is something religions should stop doing because it is a form of tribalism that increases conflict.

36. Survey by Pew Research Center for People and the Press, July 2009.

37. In *Rocks of Ages,* Stephen Jay Gould states: "Progress cannot be impeded indefinitely, and if theology does not yield its former control over the proper magisterium of science, then religion, with all its virtues, will die."

38. Shiri Noy and Tomothy O'Brien, "Traditional, Modern, and Post-Secular Perspectives on Science and Religion in the United States," *American Sociological Review* 80, No. 1 (2015): 92–115.

39. Robert Putnam and David E. Campbell, *American Grace: How Religion Divides and Unites Us* (New York: Simon & Schuster, 2012), 112.

40. Jack Good, pastor (retired), The United Church of Christ, http://www.the-brights.net/vision/essays/dennett_good.html.

41. Any religion can be modified to strip away elements that are inconsistent with generally accepted scientific fact and have left the values that are not inconsistent. For an example of such a modification of Christianity, see Gretta Vosper, *With or Without God: Why the Way We Live Is More Important Than What We Believe* (Toronto: HarperCollins Canada, 2008). For an example of such a modification of Buddhism, see Ian Flanagan, *The Bodhisattva's Brain: Buddhism Naturalized* (2011).

42. Jared Diamond, *The World Until Yesterday*, ch. 9. The definition of religion for legal purposes is especially problematic. In *The Impossibility of Religious Freedom* (2007), W. F. Sullivan persuasively shows that it is impossible for a legal system to determine fairly what is religion and what is not. Consequently, laws that try to treat religion differently from nonreligion are and always will be inherently unfair. This is one of the reasons why we argue that any claim that a person or position is "religious" should be respected, but that legal rights that attach to being "religious" should be minimal or zero.

43. Some authors who have given this issue much thought strongly disagree with us. For a detailed explanation, read the piece co-authored by Tom Flynn, Ronald A. Lindsay, and Nicholas J. Little published as the lead editorial in the February/March 2015 issue of *Free Inquiry*. http://www.centerforinquiry.net/blogs/entry/secular_humanism_not_a_religion/. We understand these authors do not want their organization or congregations to be considered "religious" for reasons based on history: religions have caused a lot of distress for people who are secular or atheists or evidists and they do not want to be associated. However, their approach perpetuates unhelpful tribalism—the religious tribes against the nonreligious tribes. Under their

approach, the only way to end this tribalistic divide is to persuade all people to leave their "religious" congregations and then, if they want a congregation, join a "nonreligious" congregation. It will be much easier to persuade individuals to change their congregations from within to be secular and evidist. To do this, we must define the word "religion" broadly enough to encompass all humanist groups, and *any* group, leaving each group latitude to call itself "religious" or not. Although Flynn and his colleagues vehemently do not want to "come to religion," we propose that, instead, "religion" should come to them. Instead of a tribalistic divide between the religious and the nonreligious, which is a vague dividing line because there is no agreement on the meaning of the word "religion" and never will be a consistent definition of the word, we will then have a tribalistic divide between evidists (both religious and nonreligious) and nonevidists. This is a good divide to have. There is little doubt where the dividing line lies, and, for culture to move forward, it is important to highlight this dividing line. In addition, there are already evidist groups that call themselves "religious" and we will not be able to persuade them to stop doing so, in part because they get benefits of religious privilege laws when they do so and they will not want to give up these benefits.

44. For example, some religions claim a privilege under the U.S. Constitution to keep and handle dangerous poisonous snakes despite wildlife-control laws and risk of death to members of the congregation and others. https://en.wikipedia.org/wiki/Snake_handling#Legal_issues. These positions receive little support in the United States. One could argue that such practices should be allowed on grounds of "religious freedom" if they are limited to private homes so there is insignificant effect on others.

45. No privilege of any consequence should be accorded to a religion-based position compared to a secular position. For example, each social club like the Sierra Club, the Seattle Mountaineers, Kiwanis, or the Elks should be entitled to call itself a "religion" if it wishes to do so and should be entitled to all the tax benefits and other privileges. If large numbers of groups win the privileges, political pressure will build to eliminate those privileges.

46. In *Rocks of Ages,* Stephen Jay Gould uses the word "religion" to encompass all exploration of ethics, morality, values, and meaning, including by atheists. We agree that the word *religion* should be defined this broadly. Some argue that allowing a group to call itself "religious" has social and legal implications, so we must carefully guard this gate and only allow selected groups to pass. Our position is that (a) there is no consistent or principled way to decide what is a religion and what is not, and (b) the legal implications should be eliminated.

47. The Sunday Assembly is a growing set of congregations across the

United States that meet on Sundays and provide an alternative to "religious" congregations. This group is not included in this section because, although it embraces some elements of traditional religious observance (weekly meetings, music, inspiration), the movement does not refer to itself as "religious."

48. William R. Murry, *Reason and Reverence: Religious Humanism for the 21st Century* (2010) http://www.uua.org/publications/skinnerhouse/browseskinner/titles/14627.shtml; *Becoming More Fully Human: Religious Humanism as a Way of Life* (2011) http://thehumanist.org/september-october-2013/becoming-more-fully-human-religious-humanism-as-a-way-of-life/#sthash.ZfYiU04E.dpuf.

49. "Religious Humanism Today," *Free Inquiry* (October 2013) (emphasis added) https://www.secularhumanism.org/index.php/articles/3561.

50. Holden et al. v. U.S. Federal Bureau of Prisons et al., http://americanhumanist.org/system/storage/2/b7/c/5080/20140408_Oregon_Complaint.pdf.

51. Holden et al. v. U.S. Federal Bureau of Prisons et al., http://americanhumanist.org/system/storage/2/fb/9/5632/Motion_to_Dismiss_Denied_10-30-14.pdf.

52. Holden et al. v. U.S. Federal Bureau of Prisons et al., http://americanhumanist.org/system/storage/2/52/9/5666/AHA_v._BOP_Final_Settlement_All_Signatures.

53. For the values of the Unitarian Universalist Association, see note 27 in this chapter.

54. Estimate by William R. Murry, 2013.

55. Bill Maxwell, "Leading the Unitarian Universalist Association, a Faith without a Creed," *St. Petersburg Times* (April 11, 2008).

56. Center for Pragmatic Buddhism, http://www.pragmaticbuddhism.org/node/12. It appears they find "rebirth" to be a useful concept for determining values and not a fact. http://www.pragmaticbuddhism.org/philosophy.

57. Center for Pragmatic Buddhism, *A Pragmatic View of Religion* (2013); http://engageddharma.com/2013/06/18/a-pragmatic-view-of-religion/.

58. http://www.shj.org/FAQs.html.

59. http://www.shj.org/FAQs.html. The congregations, communities, and *havurot* (small groups) are affiliated with the Society for Humanistic

Judaism, a North American "religious" organization.

60. http://ethicalfocus.org/how-is-ethical-culture-religious/.

61. Arthur Dobrin, quoted in Jone Johnson Lewis, "Ethical Culture as Religion," 2003, American Ethical Union Library.

62. This and other aspects of Quaker belief and practice are spelled out at "Meeting the Spirit: An Introduction to Quaker Beliefs and Practices," http://charlestonwv.quaker.org/meeting-the-spirit.html.

63. http://en.wikipedia.org/wiki/Lloyd_Geering.

64. http://www.religiousconsultation.org/.

65. Retired American bishop of the Episcopal Church. See, e.g., *A New Christianity for a New World: Why Traditional Faith Is Dying and How a New Faith Is Being Born* (2002).

66. Vosper, *With or Without God*.

67. Tom McLeish, *Faith and Wisdom in Science* (Oxford: Oxford University Press, 2014; Kindle Edition).

68. McLeish, *Faith and Wisdom in Science* (Kindle Edition location 4992).

69. McLeish, *Faith and Wisdom in Science* (Kindle Edition location 3988).

70. "Going the Whole Hog," *Times Higher Education*, 1995. http://www.timeshighereducation.co.uk/story.asp?storyCode=97718&sectioncode=26.

71. http://www.samharris.org/site/full_text/what-barack-obama-could-not-and-should-not-say#sthash.iTqbNntC.dpuf.

72. http://www.samharris.org/site/full_text/the-strange-case-of-francis-collins#sthash.owReuELp.dpuf.

73. http://www.samharris.org/site/full_text/response-to-my-fellow-atheists#sthash.yohRTEIw.dpuf.

74. Video formerly posted on YouTube, no longer available.

75. Christopher Hitchens, *Letters to a Young Contrarian* (New York: Basic Books, 2001), 55.

76. Daniel Dennett, *Darwin's Dangerous Idea* (New York: Simon & Schuster, 1995).

77. Alex Rosenberg, *The Atheist's Guide to Reality* (New York: W. W. Norton, 2011), 111.

78. "Seeing and Believing," *New Republic*, 2009.

79. Daniel C. Dennett and Linda LaScola, *Caught in the Pulpit: Leaving Belief Behind*, 2nd ed. (Durham, NC: Pitchstone Publishing, 2015; Kindle location 1230).

80. Richard Dawkins, *The God Delusion*; http://en.wikipedia.org/wiki/Spectrum_of_theistic_probability.

81. Those who, without believing in any god, follow Pascal's wager and pray to a god on the chance that he or she or it might exist, if there are any such people, are fairly called agnostics but not atheists because they are basing some actions on the chance that there might be a god. Lex Bayer and John Figdor explain that those who choose to emphasize the near-zero probability that a god exists characterize themselves as atheists and those who choose to emphasize that the probability is significantly greater than zero characterize themselves as agnostics. *Atheist Mind, Humanist Heart* (Lanham, MD: Rowman & Littlefield, 2014), 10.

82. Before the introduction of modern religions in Madagascar, first Islam by Arabs and then Christianity by Europeans, the most salient religious practice was supplication of your personal ancestors for guidance or help. This included elaborate death rituals for dead parents and grandparents and placing money and valuable goods at sacred sites as gifts to one's personal ancestors. Given any useful meaning of the word "god," we query whether this practice should be considered "theistic" or "atheistic." Should we consider all the dead ancestors to be "gods"?

83. Reza Aslan, http://www.salon.com/2014/11/21/reza_aslan_sam_harris_and_new_atheists_arent_new_arent_even_atheists/.

84. Beth K. Lamont with Beverley Earle, "Introduction" to *The Philosophy of Humanism* by Corliss Lamont, 8th ed. (Washington, DC: Humanist Press, 1997).

85. Steven Pinker says that "the acquisition of knowledge is *hard*. The world does not go out of its way to reveal its workings, and even if it did, our minds are prone to illusions, fallacies, and superstitions." Science Is Not Your Enemy: An Impassioned Plea to Neglected Novelists, Embattled Professors, and Tenure-less Historians http://www.newrepublic.com/article/114127/science-not-enemy-humanities#.

# 4

# HOW TO MARKET
# SECULARISM AND SCIENCE

The sixteen suggestions in chapter 3 can be helpful for anyone who has not yet fully accepted the scientific consensus on all facts (evidism) and the importance of secularism in all public institutions. They also offer guidance for leaders of the secular movement. The next nine suggestions are intended mostly for secular leaders and authors.

## USE THE WORD SECULAR IN A WAY THAT ALLOWS PEOPLE WHO CALL THEMSELVES RELIGIOUS TO BE SECULAR AS WELL.

As discussed above, various authors ascribe different meanings to the word *secular*. In this book, we argue that it will help cultural evolution move forward if all those who consider themselves to be religious become *secular* within the core values meaning, as well as *evidist*. As we use the terms, *secular* and *evidist* have distinct and complementary meanings. People can be secular and not evidist. They can be secular and traditionally religious.

Setting aside for a moment the benefits of evidism, if we could move all people to accept secularism, it would greatly benefit mankind. As discussed above, it is the only practical way to maintain the peace. Only the core values meaning of secularism is required to obtain these benefits. The concepts that people must understand to obtain these benefits are relatively simple and clear. We need a word to represent these concepts so that the concepts can be clearly promoted. At this time, that word is *secular*. There is no synonym. Large numbers of religious people have supported the use of this word for these concepts since George Jacob Holyoake coined it in 1851.[1] If the meaning of the word is now taken in another direction, our ability to promote secularism will be damaged.[2]

The meaning that some newer authors are giving to the word *secular* is synonymous with *nonreligious* and close to *secular plus atheist*[3] or perhaps *secular plus evidist*. Whatever benefit might come from having a word for these conjoined ideas is not worth the loss to our ability to spread the core concepts of secularism. In this book, we argue that we need a word for evidism. Merging the concept of evidism into other concepts such as secularism will make it harder to teach and explain evidism itself. The word *evidism* works well as an adjunct to the word *secularism*. Once people understand and accept secularism, they are primed to understand evidism.

Furthermore, these authors are giving the word *secular* content and meaning only in contrast to *religious*. They are saying, if you are not religious, then you are secular; you cannot identify as both religious and secular. This gives the word *secular* no clear meaning. Thought leaders do not agree on the content of what should be considered *religious*, in part because the word has never had a clear meaning and in part because the scope of possible meanings has been changing over time. Indeed, we are arguing that the scope should be made so broad as to have no clear content.

These authors should stop using the word *secular* to mean nonreligious wherever they are implying "you cannot identify *secular* and *religious* at the same time." In these situations, they should just use the word *nonreligious* instead.

## WHEN ADVOCATING SECULARISM AND EVIDISM, AVOID PUBLICLY ARGUING FOR IRRELEVANT VALUES ON WHICH PEOPLE DISAGREE.

Because values cannot be determined by science, secular leaders cannot say that people are objectively wrong when they espouse particular values, no matter how abhorrent those values may be, unless they violate principles of secularism. If we claim that a person's values are wrong, that person will likely stop listening. This is a good reason to avoid talking about values on which people disagree when trying to persuade others to be secular and evidists. But we can say people are objectively wrong when they espouse wrong facts. If we claim that a person's facts are wrong when they are inconsistent with science, we will win the argument sooner or later, provided the person doesn't stop listening and we don't stop presenting the evidence.

As discussed above, for optimal connection with an audience, the only four values that narrowly focused advocates for secularism and evidism should talk publicly about are the value of evidism for all aspects of life and the three values essential to secularism:

1. **Equality.** Equality of all tribal/ethnic/religious groups in all aspects of government, including freedom of speech on topics of facts, values, and religion—no apostasy, heresy, or blasphemy laws.

2. **Liberty.** Freedom of conscience not to be coerced to take actions or refrain from actions, including religious practices, that violate one's conscience, unless required to avoid harm to others; tolerance of diverse and conflicting values, freedom of speech, and privacy.

3. **Truth in Government.** Base all government decisions on the scientific way of knowing—discovering objective facts and relying on objective facts.

We are not suggesting that secular advocates should claim to have no values other than evidism and the three core values of secularism. To maintain their persuasiveness, narrowly focused secular advocates should simply decline to talk publicly about other particular values on which people disagree. And secular advocates should also decline to talk publicly about their religion (the content of which should be confined to values if they are evidists). On the topic of your values, you could say publicly something vague like: "Other than the values of secularism and the scientific way of knowing (evidism), my values are essentially the ones shared by people of many perspectives, including Christians, Jews, Muslims, Buddhists and humanists."

When debating values, people often assert false "facts." It is good to show that their facts are wrong, but if you want to maintain maximum effectiveness with a public audience, don't make the leap to asserting that their values are wrong. The farthest you should go is to suggest that they might want to reconsider their values once they realize that the facts they were asserting to support those values are wrong.

## WHEN MARKETING THE BEST WAY OF KNOWING FACTS, USE AN EFFECTIVE LABEL.

Scientists and philosophers have reached a consensus that there is a best way of knowing all facts, but they have not yet agreed on what to call it. Some call it *empiricism,* but others argue this label is not broad enough.[4] If we want to spread this way of knowing, we need a label that is easy to understand, one that everyone will be comfortable using. *Empiricism* isn't easy enough for people to say or learn, so it is not effective enough for marketing. It also doesn't help that there is still some philosophical debate about the meaning.

A candidate for this label is *the scientific way of knowing.*[5] This label includes no unfamiliar words and calls on everyone either to confirm that they accept the scientific way of knowing or to concede that they reject some of the conclusions of science, which can encourage them to re-examine their views. This label doesn't imply that all valid knowledge comes from science; it only implies a deep skepticism of knowledge claims that are inconsistent with scientific consensus and a cursory dismissal of such claims except when undertaking scientific or philosophical investigations.

## ADOPT A MARKETABLE LABEL FOR PEOPLE WHO ACCEPT THE SCIENTIFIC WAY OF KNOWING.

It would be helpful to have terminology to refer to people who accept the scientific way of knowing—terms that can be easily understood and used by everyone, not just academics. The new terms that would be helpful include:

1. a *person-categorizing noun* for adherents of the scientific way of knowing—words that typically end in -ist or -ian or -an;

2. an *adjective* for characterizing the scientific way of knowing—words that typically end in –ic or –istic or –ial or -al; and

3. a *school of thought labeling noun* to identify the scientific way of knowing—words that typically end in -ism.

Some argue that an "ism" noun is not needed for the scientific way of knowing. Some even feel it is counterproductive—that it misleads as to the true nature of the scientific way of knowing because it shouldn't be a theory or "ism" at all. The reasoning goes that theories can become bound with dogma—attempts to characterize the theory with a particular set of words—which is inconsistent with the self-correcting nature of the scientific spirit.[6] We don't discount those arguments. But it is a practical reality that "ism" nouns for the scientific way of knowing are already in use, including *naturalism, scientific rationalism,*[7] and *scientism.*[8] If our marketing and meme spreading is to be effective, we should provide people with preferred terminology that they will be persuaded to use in place of their current usage.

## *THE INADEQUACY OF CURRENT LABELS*

It is possible to give too much emphasis to words and labels, to spend so much time on what we call something that we never get around to the substance those labels represent. Still, a label that brings up the wrong associations for a listener can get in the way of productive discussion. This is especially problematic if the difference in understanding isn't made clear. Most of us have had the experience of getting ten minutes into an argument, only to find that the other person was operating with a different understanding of the definition of a vital word—like *spirituality*, or *atheist*, or even *science*.

This section looks at the shortcomings of existing alternative labels for the scientific way of knowing.

### *Naturalist / Naturalistic*

Some suggest we should use the adjective *naturalistic* to identify the scientific way of knowing because it is based only on good evidence from the natural world,[9] and *naturalist* for the personal noun. This meaning of *naturalism* has decades of momentum, but there hasn't been much headway outside the academic world.

As noted above, most people immediately think of a naturalist as someone who studies life forms or explains nature to public audiences. It would be virtually impossible to re-educate the public in a new definition of the word.

And philosophical writing uses the word *naturalistic* in confusing and inconsistent ways. Some authors use it to denote the scientific way of knowing the facts, while for others it means a philosophy without supernatural concepts that draws supposedly objective conclusions about values from observations of nature.[10]

Furthermore, proponents of many antiscientific theories, such as homeopathy and religious naturalism,[11] claim that they are describing nature and there is nothing supernatural about their theories. We can tell them that they are wrong and that their theories invoke supernatural forces, but they will not accept this characterization. If they will not accept the label we apply to them, the label loses most if not all of its utility. The problem is that the proponents of these theories don't view generally accepted scientific methods as valid for testing their theories.

The use of *supernatural* and *naturalistic* place too much emphasis on the desire to end the influence of belief in a god or spirits and not enough emphasis on ending other beliefs that are inconsistent with scientific consensus. They grew out of the conflict between theism and atheism, which will lose importance as theism continues to fade. And while belief in gods may still be the strongest obstacle to a shared way of knowing in human culture, it is not the only obstacle. Adopting a way of knowing that includes no gods but still does not reflect the importance of accepting the scientific consensus on all facts fails to take people across the bright line into the scientific way of knowing. So the labels *supernatural* and *naturalistic* miss the opportunity to cast helpful light on other failings of the prescientific, intuitive, traditional, natural ways of knowing.

### Bright

In 2003, an association of people free of supernatural or mystical beliefs organized an effort to promote *bright* as a new term to describe people with a naturalistic worldview.[12] It has many advantages, including that it is short, easy to say and spell, upbeat, and memorable. The philosopher Daniel Dennett, a supporter of the word, later suggested the contrasting label *super* for those who maintain beliefs in the supernatural.

But two problems doomed the effort. The word *bright* was quickly derided as too self-congratulatory, especially when the antonym (*dim*) is applied to others. There is also the problem that the word *super* will never be widely accepted by those it is intended to characterize, and such acceptance is important to making progress.

While the Brights' effort is a laudable attempt to solve the same marketing problem we are addressing here, the failure of this terminology to gain serious traction[13] shows that it is time to try again.[14]

### Skeptic / Skeptical

Many people use the words *skeptic* and *skeptical* to identify the

scientific way of knowing.[15] And here again we run into a serious conflict between the popular understanding of a word and the needed meaning.

The popular image of a skeptic is a curmudgeon who doubts for the sake of doubting and refuses to consider any possibility that isn't vividly clear. This fails to capture the affirmative confidence of the scientific way of knowing that there is no part of reality that cannot be discovered through inquiry consistent with science.[16] And many people who call themselves skeptics are not evidists; for example, those who deny the evidence of climate change caused by humans.

So once again we run into the problem of a term with multiple meanings,[17] some of which are inconsistent with the scientific way of knowing. We need a short, single word that is not burdened with other meanings. Such a word will give us a concise, clear way to respond to any assertion by saying, "I am a _____." If the listener knows what _____ means, the response will be understood. If the listener doesn't know what the term means, they will ask what it means *instead of assuming a meaning that isn't true*. *Skeptic* will not work for this purpose because the pre-existing definition leaps into the listener's head, and no opportunity is presented to clarify. Until the new meaning of skeptic is widely understood and other meanings disappear from lack of use, people will assume the speaker merely means to say they doubt a given assertion, not that they are trying to convey the value of the scientific way of knowing.

*Scientist / Scientific*
We could take the words we need from the word *science,* as we propose for "the scientific way of knowing." The "ism" would be *scientism*. Despite efforts by some critics of scientific thinking to use "scientism" as a derogatory term, these words have suitable connotations and there are advocates for this terminology.[18]

It is important to also have a person-categorizing noun, but there is no good existing candidate for this. The obvious choice would be *scientist* or *Scientist,* but we instantly run headlong into the common usage of the term. Most people understand a *scientist* to be someone whose professional work uses scientific methods, or a person with a specific educational degree. We would need to change the understood meaning of *Scientist* to refer to an understanding that anyone can have, not just those who make their living or hold a degree in the sciences.

There is also the important issue of self-definition. There is no errand more foolish than trying to force a new label on people who don't accept it themselves. One United States study found that many people who accept the scientific way of knowing are unwilling to call themselves "scientists."[19] And, as the word "science" is understood in English, it excludes fields such as mathematics and history—both valid sources of knowledge that share at least some methodology with the physical sciences.

In addition, some of the people who self-identify as scientists and are accepted as such by all sectors of society employ the scientific way of knowing only in their work and not in their personal thoughts and lives 24/7. For example, Francis Collins, head of the National Institutes of Health, says he believes in a god and has published books explaining his strongly supernatural views.[20] It would torture the meaning of the word to claim that Collins is not a *scientist*, but his expressed views are clearly inconsistent with the scientific way of knowing.

So the game is over for any useful redefinition of *science* or *scientist* before it even begins. The obstacles are insurmountable.

*Atheist / Atheistic*

Some people use the words *atheist* and *atheistic* to identify the scientific way of knowing. The biologist J. B. S. Haldane identified atheism as a necessary precursor to his work. "My practice as a scientist is atheistic," he wrote. "That is to say, when I set up an experiment I assume that no god, angel or devil is going to interfere with its course; and this assumption has been justified by such success as I have achieved in my professional career. I should therefore be intellectually dishonest if I were not also atheistic in the affairs of the world."[21]

But even if we accept this as a condition of scientific work, it cannot be made equivalent. This use of *atheism* would be confusing to audiences for whom it means merely the lack of belief in a god. Yet speakers who use it in this way often intend to convey much more than this, including acceptance of the entire scientific way of knowing. The word atheist inadequately conveys this more comprehensive sense, and it would torture the language to try to redefine what *atheist* means to fill the need. The word *atheist* already has a job, and it does that job well, providing a useful contrast with *theism*. It would diminish the usefulness of this word to stretch it to refer to all aspects of the scientific way of knowing.

There is also the same issue of inconsistency discussed above for scientists: there are atheists who hold factual beliefs that are inconsistent with scientific consensus, such as homeopathy, or astrology, or that nature (all life on Earth) is a valid source of objective values,[22] or that objective values can be derived from a source outside of humanity.[23]

*Secular / Secularist*

Finally, there are authors that choose to use the words *secular* and *secularist* to refer to people who reject beliefs in anything supernatural or transcendent.[24] This terminology will not work to replace *naturalism* because, as discussed above, this meaning of *secularism* is quite different from *naturalism* and there are other meanings of *secular* that are important concepts and need this word to have a specific meaning.

## CRITERIA FOR SELECTING NEW TERMINOLOGY

One of our targets for marketing the scientific way of knowing should be people who are causing harm to others and using false factual beliefs for their motivation or justification, such as religious extremists. As we choose better labels, we should keep in mind whether the labels will be optimally effective for reaching these people.

The target where we may achieve the most success and therefore make the biggest difference for the evolution of culture is the "Nones,"[25] as well as the supernaturalists who might switch to Nones with a little encouragement. Researchers report that large numbers of Nones have no interest in either theism or atheism.[26] The new terminology should make no implicit reference to the long-standing theism/atheism debate. The day will come when this issue is considered by most to be inconsequential, and planting a flag on one side or the other will only serve as an unneeded obstacle at this point. We need terminology that will be useful for making progress on the rest of what is important to bring each person to accept the scientific consensus on the facts.

The words we choose should work well in translation to all important languages. For example, before expanding into Germany, a company called GiftCo might want to know that the word *gift* in English means *poison* in German. Likewise, a salsa distributor aiming at the South Korean market should be aware that *salsa* is almost identical to the Korean word for *diarrhea*. If our new terminology is to have maximum utility around the world, we would do well to avoid similarly unfortunate linguistic slips.

In English and Romance languages, the "ism" word, the "istic" word, and the "ist" word should have a single root with three varied endings to make them easy to learn. The root should convey desired meanings well so it is easier to teach the words. If we choose a root that is Latin or Greek, the same root is likely to work in many languages. Ideally, the root would require no spelling change for most other languages. It would also be best if a connotation of the root helps people distinguish between conclusions based on facts and conclusions based on values. It would be helpful if the root creates a mental association with one of the commonly used words *science* or *evidence* or *reason* or *reality*.

An important question is whether we should use a word that already has a suitable general meaning, such as *empirical* or *evidential*, and educate everyone to a new, more specific meaning for this word, or whether we should coin a new word.

Here is an argument against using an existing word with a suitable general meaning: a person or group that does not adhere to the scientific way of knowing facts and yet applies the imprimatur of this word to their views will cause confusion as to what the word means. For example, if we choose *scientific, evidential, empirical, naturalistic,* or *skeptical,* a person promoting a view that is inconsistent with the scientific way of knowing might use these words with a different meaning, and confusion would ensue.[27]

To avoid this problem, we need to coin a new word and overpower with public speaking and publications any effort by others to change the meaning of the new word in a wrong direction.

To coin a new word, we can compose a root that is a novel string of letters with a novel sound, or we can select an existing word with no related meaning, such as *rainbow* or *apple*. Selecting an existing word with no related meaning will make the education task more difficult than composing a novel string of letters that creates a helpful mental association.

The best balance of considerations is to compose a new root with enough resemblance to an appropriate word and no undesirable associations. Perhaps we can compose a new root that sounds like one of the roots in the commonly used words *science, evidence, knowledge, reason,* or *reality*.

## PROPOSED NEW LABELS: *EVIDIST* / *EVIDISM*

For all the reasons given above, the founders of the Brights were right to select a new word. They unfortunately chose a word with fatal flaws. In the scientific way of knowing, beliefs are consistent with evidence and reason validated through the peer-reviewed scientific process. Referring to the scientific way of knowing with the adjective *evidential* or the adjective *empirical* has appropriate connotations. Unfortunately, as explained above, if we choose one of these words with a suitable general meaning, people with other ways of knowing can easily co-opt them for their own purposes.

Instead, we can use the sound of "evidence" to coin a new root and make the needed new words. We propose *evidist, evidism,* and *evidistic.*[28]

In English, these words trigger an appropriate association with *evidence.* We do not believe they will trigger any undesirable associations for speakers of English. Because *eviden* is a Latin root, it is likely that *evidist* will have the same association in all Romance (Latin-based) languages that it has in English. It would be best if the same spelling of *evidist* will also work in all other languages that use the Latin alphabet.[29]

There is a passing concern that some who do not value the scientific way of knowing might feel denigrated by the term's implication that they really should value it. There is probably no avoiding this, no matter what term is used. The term *bright* was often found insulting because it seemed to imply an off-topic insult (that others are not intelligent). If *evidist* is considered insulting, it would at least be an on-topic criticism. Insulting people over their failure to understand what is good evidence may be worth the drawback because this is exactly the point we most want to make.

As people adopt evidism as an overriding amendment to the beliefs and values provided by their preferred religion, most will at first keep their views secret from friends and family. When these people come out to friends and family they can adopt labels like: "I am an evidist Muslim—I am a Christian evidist—I am an evidistic Buddhist—I am a Sunni evidist—I am an evidistic Shia—I am an evidistic Hindu—I am an evidistic Unitarian—I am an evidistic humanist."

## UPDATE THE COMMONLY USED DEFINITIONS OF RELIGION.

In 2012, Jared Diamond listed sixteen published definitions of religion. He found each imperfect for modern usage and proposed his own, a seventeenth.[30] He observed that the functions of religion have changed over time and that, for this reason, the most useful definitions have been changing over time.

Carefully articulating definitions can help us usefully organize our models of reality and keep concepts separate that should be separate. To keep the definition of "religion" most useful, Diamond does not consider water-witching, belief in magic, patriotism, Confucianism, the core of Buddhism, or philosophy of life to be religions. He noted that Confucianism and the core of Buddhism are considered by others to be both philosophies of life *and religions*. Like Diamond, the Australian philosopher Russell Blackford asserts that in understanding human nature and cultures, it is most useful to have a word for that which people consider to be sacred or transcendent. Blackford claims that limiting the use of "religion" to that meaning is the present best choice for the word.

However, none of the six groups of people listed in chapter 3 that call themselves religious and yet fully accept the scientific consensus on the facts fit into any of the seventeen definitions. As discussed above, it is unwise for marketing reasons to define the word *religion* so narrowly as to exclude any groups that call themselves *religious*. Where the intended meaning is limited to religions that include something supernatural or transcendent or sacred, we could call them "nonevidist or supernaturalist religions" or use similar adjectives.

Of the seventeen proposed definitions listed by Diamond, here are five that, with minor amendments as shown with strikethrough and underlining below, are broad enough to encompass such religions. "Belief" is to be understood as encompassing belief in the scientific way of knowing all facts and beliefs about values.

1.  Webster's New World Dictionary: "Any system of belief ~~and worship~~, often involving <u>worship,</u> a code of ethics and a philosophy."

2. Lessa and Vogt: "A system of beliefs and practices directed toward the ~~ultimate~~ important value concerns of a society."

3. William Irons: "a belief that the highest good is defined by ~~an unseen order~~ a set of values, often combined with an array of symbols that assist individuals and groups in ordering their lives in harmony with this ~~order~~ set of values and an emotional commitment to achieving that harmony."

4. Emile Durkheim: "a unified system of beliefs and practices relative to ~~sacred~~ important things or values ~~that is to say, things set apart and forbidden~~—beliefs and practices which unite into one single moral community called a Church, all those who adhere to them."

5. Michael Shermer: "a social institution that evolved as an integral mechanism of human culture to create and promote myths or values, to encourage altruism and reciprocal altruism, and to reveal the level of commitment to cooperate and reciprocate among members of the community."

Lloyd Geering's favorite definition of religion is that of Carlo della Casa[31]: "Religion is a total mode of the interpreting and living of life." This definition successfully avoids relying on particular facts and puts the focus on human values.

To avoid holding back the continued favorable evolution of religions, authors and leaders should use the updated definitions of *religion* listed above.

## ADOPT AN EFFECTIVE LABEL FOR THE TRADITIONAL WAYS OF KNOWING.

For use in our marketing of the scientific way of knowing facts (evidism), it would be helpful to have effective contrasting labels for the traditional ways of knowing. Perhaps it would be best to call the old, naturally evolved ways of knowing based on human nature *traditional* or *intuitive* or *prescientific* ways of knowing.

It would also be a fitting use of the language to refer to the traditional ways of knowing as *natural,* but this would bring us back into conflict with existing meanings and cause confusion.

Others have called the naturally evolved ways of knowing *supernatural* because they include belief in supernatural spirits. However, the term *supernatural* is too narrow because the word fails to encompass beliefs that are just as unreliable as supernatural beliefs where the proponents of these believes do not agree that they are based on anything supernatural. There are many people today who claim to have no beliefs in spirits or anything supernatural or mystical, yet they have beliefs in false theories such as fate, karma, new-ageism, homeopathy, astrology, unlucky days, or values from a source other than humans, which shows that their way of knowing is still inconsistent with the scientific way of knowing.

*Intuitive* and *traditional* appear to be our best options for labels to refer to any of the naturally evolved, nonscientific ways of knowing. A person who continues to follow a naturally evolved way of knowing when shown to be false by scientific consensus we could call an *intuitivist* or a *traditionalist*. Those who accept intuitive or traditional beliefs about facts that are wrong might find each of these labels helpfully inoffensive.

## BE POLITICALLY ACTIVE TO ACHIEVE GOOD EDUCATION IN PUBLIC SCHOOLS.

As discussed above, secularism protects freedom to practice and advocate any religion, and this includes freedom to teach false "facts" to children. Generally, secularism requires that the state refrain from telling adults that their religious facts are wrong or telling children that their parents are teaching them false "facts." But the state has a strong interest in the education of both children and adults and teaching them all the conclusions of science. Should the state refrain from teaching facts that conflict with religious beliefs to implement principles of secularism?

Leading authors on the topic of secularism, such as Russell Blackford, argue that the state has no business teaching adults or children that facts asserted by various religions are wrong.[32] They make an exception only for countering religious teachings that are likely to cause substantial harm. For example, these authors support teaching of human evolution to children over the protests of parents because to do otherwise would cause harm to the children by leaving them scientifically ignorant.[33] These authors would allow public schools to teach that there are no unicorns or leprechauns or that Zeus is only a myth, but they would allow secularism's protection for freedom of religion to block teaching that there are no gods or souls or afterlife, arguing that allowing children to believe these myths does them little harm.

For people who value evidism—that everyone should learn the truth about reality—this theory is deeply upsetting. The state should teach all scientific truths to all adults and all children, including the truth about gods! There can be no political principles that block teaching of truth.

At the present time, there are indeed limitations on which scientific facts that conflict with religion the state may teach. The resulting outcome argued by Blackford and other authors may be correct in each particular case, but the theory asserted is wrong. Correctly interpreted, principles of secularism and freedom of religion do not answer these questions. They do not block the state from teaching any scientific truths, and they do not obligate the state to teach any scientific truths.

The relevant principle here is the value of evidism—the value of acceptance of scientific consensus on all facts. The principle of evidism conflicts with complete freedom not to have your government teach that your religious facts are wrong. How should these competing values be balanced?

The answer is provided by political practicality. In an ideal world, the state would teach all facts to all children and adults. However, so long as a majority believes any particular false "facts" asserted by religions, it is politically impossible to get the state involved in teaching against these "facts." A political majority now believes the truth of human evolution, so this can be taught. However, in 2016, the public schools in the United States cannot teach that there is no god or no afterlife or that people have no souls, because a political majority would block such teaching.

When cases presenting these conflicting values are raised in the courts, such as cases in the United States striking down laws that prohibit teaching of human evolution in public schools, the courts engage in legal sophistry with theories like the assertion that principles of secularism stop the schools from teaching facts inconsistent with religion unless children would be clearly harmed. The courts cannot openly consider what a political majority in the country might believe, but this is what the judges are really thinking. Although they can strike down a law that prohibits teaching evolution, they will not be willing to strike down a law that prohibits teaching that the god of Abraham is a myth because to do so would be exceedingly unpopular. And it may be impossible for individual judges who believe in that god and are not evidists even to conceive of the correct answer.

It is proper that the courts mask their real thinking with legal sophistry, so it is helpful to the courts and progress of human cultural evolution that authors like Blackford help them by providing analyses they can use. Until a majority accepts the principle of evidism, it will help us nevertheless move forward if authors like Blackford provide helpful legal sophistry with their interpretation of principles of secularism.

The correct analysis of principles of secularism and freedom of religion is that they place no restraint on the state teaching scientifically validated facts that conflict with religious teaching. People are free to practice their religions and teach their children any factual assertions they like, but they have no grounds to ask the state not to teach conflicting truths validated by science.

The only limits on what scientific facts the public schools can teach are political. Once a political majority no longer believes that the sun revolves around the earth, the truth can be taught in public schools. It is the same for the age of the earth, evolution of humans, the existence of hell, the existence of gods, and the existence of souls. As soon as a political majority no longer believes these things, we can teach the truth in public schools. In the meantime, political reality is holding back public education. It will be an exciting day when a political majority accepts all the conclusions of science, including that no god is more likely than leprechauns, and all conclusions of science can be taught in public schools.

Principles of secularism do not require that we refrain from lobbying for teaching by the state of all facts to both adults and children. But we must accompany that lobbying with private education of politically active adults, because we will never win the right to teach the truth without support from a political majority.

## ENCOURAGE A COMMUNITY TO MAKE THE SCIENTIFIC WAY OF KNOWING ITS NORM.

Over the three centuries since the beginning of the Enlightenment, scientists and philosophers have developed the scientific consensus on the facts to the point that it can be easily adopted by anyone who chooses to do so. The most important intellectual transition that a majority must achieve to move cultural evolution forward is from traditional ways of knowing to the scientific way of knowing (evidism). In the last 300 years, the scientific way of knowing has been fully adopted by only a small minority of people, although there has been progress.[34]

Is there any community where the scientific way of knowing (evidism) is accepted by a clear majority of the thought leaders? A university town, maybe? To answer this question, we might look for a community where expressions of beliefs inconsistent with science—homeopathy, for example—are met with social rejection rather than actual or feigned acceptance.

Six hundred years ago, people stating an opinion that the earth is flat or at the center of the universe would have been met with tolerance or acceptance, not dismissal. But now, serious suggestions that the earth is flat or that the sun revolves around the earth are met with rejection. Likewise, support for creationism instead of evolution has become socially unacceptable in most of the developed world in the last one hundred years—though much of the United States notably lags in this respect. The stated view that homosexuality is factually unnatural has become unacceptable in most parts of the United States in the last twenty years. These are all examples of favorable cultural evolution with respect to factual beliefs. People who express these factual beliefs that are now known to be wrong are met with rejection. *We have moved on from these misconceptions,* goes the message. *Time to join the evolution!*

Is there any village or town or region on Earth where all expressions of factual beliefs inconsistent with scientific consensus are met with rejection rather than acceptance? Wouldn't it be great for the evolution of human culture if we could get to this point at least somewhere on Earth?

## URGE LEGISLATORS TO EXCLUDE FROM PUBLIC BENEFIT PROGRAMS ORGANIZATIONS THAT PROMOTE SCIENCE DENIAL.

In June 2017, the U.S. Supreme Court ruled in Trinity Lutheran Church vs Comer that an organization could not be denied equal access to public benefits simply because it calls itself "religious" or pursues a "religious" mission. The state had offered to subsidize upgrades to private school playgrounds to make them safer but declined to do so for the Trinity organization because, as a part of the Evangelical Lutheran Church in America, it is "religious." The Court's decision was decried by secular activists who have argued for two hundred years with much success that no taxpayer money should be used to support "religion."

Although the analysis relied on by the Court may be faulty, the decision to base the conclusion on the First Amendment to the U.S. Constitution may be unsound, and the ruling may be wrong for other reasons (and perhaps federal courts should not intervene in such an issue to begin with), we think the Court's central argument is correct: whether an organization calls itself a "religion" or has a "religious" mission should not determine any decision by any branch of government that might be adverse to anyone.

We have argued in this book that the word "religion" should be defined broadly enough and vaguely enough that any organization that wishes to call itself "religious" can do so and no one will have adequate grounds to say they are wrong to do so. Consequently, no government decision should ever turn on whether an organization is "religious" or calls itself "religious."

What kinds of organizations should be denied taxpayer money then? We think it would be good policy to deny taxpayer money to any person or organization with a mission to promote science denial, including any assertion of fact that is inconsistent with the scientific consensus. Thus, governments would deny taxpayer money to those individuals and groups, including contemporary mainstream religions, that deny, for example, the scientific consensus on evolution, climate change, or vaccinations. Conversely, governments would not deny taxpayer money to the six evidist religions discussed in this book, or to Sunday Assembly,[35] which acts like a church but is evidistic and does not refer to itself as religious, or to evidistic organizations with values-based missions such as Confucian scholars, The American Red Cross, or Planned Parenthood.

Because the Evangelical Lutheran Church in America is an organization with a central mission to promote acceptance of facts inconsistent with the scientific consensus, it should be denied taxpayer money, even merely to make its playground safer for children. Once this church becomes evidistic, which we predict it will eventually, it should then be eligible for taxpayer money. If private schools want to receive taxpayer money for anything, they should ensure that they do not teach or otherwise promote any assertions of fact inconsistent with the scientific consensus.

## Notes

1. Berlinerblau, *How to Be Secular,* 56.

2. For a detailed practical analysis of why it is important not to require more than the core concepts for a person to be secular, see Berlinerblau, *How to Be Secular.*

3. For example, see Kitcher, *Life After Faith.*

4. In *Proving History: Bayes's Theorem and the Quest for the Historical Jesus* (Amherst, NY: Prometheus Books, 2012), Richard Carrier shows that history is an evidential science that differs from other sciences only in quantity and reliability of the data. Massimo Pigliucci argues that the best way of knowing includes knowledge from science, mathematics, logic, and phenomenological experience, and there is no one label that encompasses all these sources. http://philosophynow.org/issues/102/Are_There_Other_Ways_of_Knowing. Coel Hellier argues that all these sources are described by the label "empirical." http://coelsblog.wordpress.com/2014/05/22/defending-scientism-mathematics-is-a-part-of-science/. As reported by Pigliucci in the prior citation, Lawrence Krauss agrees with Hellier on this point.

5. In a prior article, Jeff T. Haley used the term "new, scientific worldview." http://freethinker.co.uk/2013/11/06/for-effective-marketing-we-need-a-better-label-than-naturalist-skeptic-or-bright/. We prefer the term "way of knowing" to "worldview" to emphasize that it is about factual beliefs. Many authors use the term "worldview" to compass both beliefs about facts and beliefs about values.

6. See, for example, Timothy Williamson, "What Is Naturalism?" http://www.opinionator.blogs.nytimes.com/2011/09/04/what-is-naturalism/.

7. Which the *Oxford English Dictionary* defines as: "A belief or theory that opinions and actions should be based on reason and knowledge rather than on religious belief or emotional response." Philosophers use this word to mean what we are referring to as evidism, but it has no traction outside of philosophy. It will not work as a word people can use to say "I am a rationalist Christian" or you are not a "rationalist," which seems to be an accusation that the person is irrational.

8. Alex Rosenberg proposes referring to the scientific way of knowing as "scientism" with no negative connotation, *The Atheist's Guide to Reality* (2011). Coel Hellier also favors use of this word. https://coelsblog.wordpress.com/category/scientism/. Christian apologists have already promoted a different connotation of "scientism" to attack it, and the word has a negative connotation in most of the populace, making it unsuitable for our purposes.

9. Examples include: (1) Alex Rosenberg, Why I Am a Naturalist, opinionator.blogs.nytimes.com/category/The-Stone/2011/09/17; (2) Tom Clark, who has numerous publications listed at his website http://www.naturalism.org/ and his blog at http://centerfornaturalism.blogspot.com/; (3) Kai Nielsen, *Naturalism and Religion* (2001) and *Naturalism without Foundations* (1996); (4) John R. Shook, *The Future of Naturalism* (2009); and (5) Richard Carrier, *Sense and Goodness without God: A Defense of Metaphysical Naturalism* (2005). Quoting one of these authors: "The basic

epistemic commitment undergirding naturalism is that we should stick with science, in partnership with philosophy, as the arbiter of what fundamentally exists."

10. http://en.wikipedia.org/wiki/Religious_naturalism, http://people. bu.edu/wwildman/relnat/. The proposed *naturalist* terminology is confusing because the word *nature* is useful in describing both the traditional and the scientific ways of knowing. *Natural* connotes the essence of the traditional ways of knowing much more than the essence of the scientific way of knowing because the traditional ways of knowing evolved naturally. The word *natural* has effective connotations for the scientific way of knowing only when contrasted with the word *supernatural*. In addition, the word *natural* has still more inconsistent meanings outside of philosophy, e.g., a style of film: http:// www.newrepublic.com/article/113506/pieta-post-tenebras-lux-stroller-strategy-stanley-kauffman.

11. http://en.wikipedia.org/wiki/Religious_naturalism, http://people. bu.edu/wwildman/relnat/.

12. http://www.the-brights.net/.

13. Countless articles have been published by people who know this terminology where "bright" would have fit well in the context but the author chose not to use it. This is convincing evidence that this effort will not reach its goals. For examples, see Mary C. Taylor, http://www.atheistscholar.org/ AtheistPhilosophies/Naturalism.aspx and http://www.atheistscholar.org/ Websites.aspx; Tom Clark, http://www.centerfornaturalism.org/allies_of_ naturalism.htm; Alex Rosenberg, Why I Am a Naturalist, http://opinionator. blogs.nytimes.com/category/The-Stone/2011/09/17.

14. The list of criteria adopted by the founders of the Brights movement for choosing appropriate labels is a useful guide: (1) short—easy-to-say/spell/ pronounce, (2); upbeat/positive term with pleasant connotations; (3) no baggage in and of itself; (4) meme potential (arouses curiosity and prompts discussion); (5) reminiscent of the Enlightenment; (6) opportunities for symbolism; (7) inclusive of anyone who sees self fitting the definition.

15. For examples, http://www.cfisummit.org/, http://www.utdash. com/, http://www.meetup.com/skeptics-128/.

16. People who calls themselves "skeptics" might also be epistemological nihilists (http://en.wikipedia.org/wiki/Nihilism), which is profoundly inconsistent with the scientific way of knowing.

17. http://en.wikipedia.org/wiki/Skeptic.

18. Alex Rosenberg proposes referring to this epistemology as *scientism* with no negative connotation (see *The Atheist's Guide to Reality*).

Coel Hellier also favors use of this word. https://coelsblog.wordpress.com/category/scientism/.

19. Focus group conducted by Jeff T. Haley, unpublished.

20. K. W. Giberson and F. S. Collins, *The Language of Science and Faith* (Downers Grove, IL: InterVarsity Press, 2011). In *The Language of God: A Scientist Presents Evidence for Belief* (New York: Free Press 2006), Francis Collins asserts that "science is not the only way of knowing." The next sentence gives his alternative: "The spiritual worldview provides another way of finding truth."

21. J. B. S. Haldane, *Fact and Faith* (London: Watts, 1936).

22. This is the position of Religious Naturalism. http://en.wikipedia.org/wiki/Religious_naturalism, http://people.bu.edu/wwildman/relnat/.

23. In *Religion without God* (Cambridge, MA: Harvard University Press, 2013), the late Ronald Dworkin argued that a "religious atheist" rejects all notions of god but accepts as objectively true two values: (1) "human life has *objective* meaning or importance," and (2) the universe as a whole and in all its parts "is itself sublime and something of *intrinsic* value and wonder." Science has shown that evolution gave humans a disposition to believe that such values are objective, and there may be good reasons to live as if these values are objective, but no evidence has yet been found to support a theory that they are in fact objective.

24. See, for example, Kitcher, *Life After Faith* (Kindle Location 365). This usage is inconsistent with well-established usage because, as we point out above, many people claim to be both *religious* and *secular*. Kitcher explains: "My choice is based on the need for some term that will cover the views of those who do not believe in transcendent entities. 'Atheism' will not do, since it restricts the class of supernatural entities (not all supposed supernatural beings are gods) and also requires denial rather than simple absence of belief (some are agnostics). 'Secular humanism' will not do, since many prominent contemporary atheists are, as this essay suggests, light on the humanism." *Challenges for Secularism* published online before 2016 at http://www.nordprag.org/papers/Kitcher8.pdf.

25. The term "Nones" is often used to describe people who indicate in surveys that they have no religion or do not belong to any particular religion. See, for example, Barry A. Kosmin and Ariela Keysar, with Ryan Cragun and Juhem Navarro-Rivera, "American Nones: The Profile of the No Religion Population, A Report Based on the American Religious Identification Survey 2008," Trinity College, 2009; http://commons.trincoll.edu/aris/files/2011/08/NONES_08.pdf. See also Tom W. Smith, "Counting Religious Nones and

Other Religious Measurement Issues: A Comparison of the Baylor Religion Survey and General Social Survey," GSS Methodological Report No. 110, 2007; http://publicdata.norc.org:41000/gss/documents/MTRT/MR110-Counting-Religious-Nones-and-Other-Religious-Measurement-Issues.pdf.

26. Pew Research, Religion and Public life Project, http://www.pewforum.org/2012/10/09/nones-on-the-rise/.

27. An example is the use of the words *naturalism* and *science* by Religious Naturalism, which claims to be "naturalistic" and "scientific" but maintains that objective values come from a source outside of humanity, i.e., nature. http://en.wikipedia.org/wiki/Religious_naturalism, http://people.bu.edu/wwildman/relnat/.

28. For a related but different concept, Paul Kurtz proposed the word "eupraxsophy" to refer to philosophies or life stances that fully accept the scientific consensus on all facts and go beyond that to emphasize the importance of living an ethical and exuberant life with particular ethical values. http://en.wikipedia.org/wiki/Paul_Kurtz#Eupraxsophy. This proposed word does not mean the same thing as *evidism* because it goes beyond acceptance of science to determine all facts and includes acceptance of particular values.

29. We invite fluent speakers of other languages to publish or send us their opinions (1) whether *evidist* will trigger undesirable associations in another language, (2) whether the same spelling of *evidist* will also work in the other language, and (3) whether *evidal* or *evidistic* works better as an adjective in the other language. E-mail jeff@haley.net.

30. Jared Diamond, *The World Until Yesterday*, 368. He does not claim that his own definition of *religion* is less tortured than the others but asserts that it corresponds well to the religions of the last 5000 years.

31. Italian scholar (1925–2014). https://it.wikipedia.org/wiki/Carlo_Della_Casa.

32. "The state "should not consider the truth or falsity of religious ideas. Religious freedom requires that the state abstain from deciding which religion, if any, is correct. . . . the might of the state should not be used to impose, suppress, endorse, promote, or disparage any religious view . . ." "The state should not be acting on the basis that it considers religious doctrines to be fantasies, falsehoods, or absurdities." Russell Blackford, *Freedom of Religion and the Secular State,* 118, 145.

33. Blackford, *Freedom of Religion and the Secular State,* 156–57). "No child has a good scientific grounding unless she understands the basics of evolution. From a religion-blind viewpoint, then, the state should teach

evolution in its public schools, and perhaps even require it in private schooling. It would do so if religion did not exist at all. The key value here is the welfare of children in a modern world that increasingly depends on science, and where scientific illiteracy closes off many careers."

34. Steven Pinker, "To See Humans' Progress, Zoom Out," *New York Times* (February 26, 2012) http://www.nytimes.com/roomfordebate/2012/02/26/are-people-getting-dumber/zoom-out-and-youll-see-people-are-improving.

35. See note 47 at the end of chapter 3.

# CONCLUSION

# A CALL TO ACTION

This book attempts to describe both the why and how of effectively sharing reality and the scientific way of knowing with friends, family, and the world. As this idea spreads, each religion will evolve and change its positions on facts to be consistent with all science, and this will reduce conflict between religious groups and between religion and science.

This is a passion project for evidists everywhere, but it is also a pressing practical concern for a world in urgent need of all hands on deck. A wide range of crises—from feeding an exploding global population to containing epidemic diseases to mitigating the effects of climate change—presents a growing threat to our very survival as a species. If solutions exist, they will be scientific ones, and their implementation will be possible only by marshaling massive resources and political will. The efforts of governments, corporations, and other collective human enterprises will be required. In most cases, these organizations act only when incented by their stakeholders—voters, citizens, shareholders, donors, boards of directors, members of religions—the many human individuals whose will drives the action or inaction of these organizations. That's you and us, and everyone we know, and everyone we don't know.

If those stakeholders are unconvinced of the value of the scientific way of knowing, they will find themselves disinclined to urge their organizations to act on that knowledge, especially if such action costs money or other resources. In the absence of that urging, the organizations with the power to address these issues have no incentive to move beyond their inertia. The solutions we already know will not be implemented, and solutions not yet found will not be found. We evidists who recognize both the value and urgency of embracing the scientific way of knowing bear a practical and even moral responsibility to find the most effective ways possible to bring others and all religions into that understanding.

Everything we know about human psychology tells us that such a project will not succeed without a sensitivity to the listener's state of mind and an effort to move the listener beyond the unhelpful biases of their various tribes. It will not succeed if religion cannot be more firmly separated from government. It will not succeed if we shun science in favor of traditional, natural, and intuitive ways of knowing. And it will not succeed so long as myths and misunderstandings about science and scientific thinking are allowed to persist.

The development of language has been one of the premier drivers of human progress in cultural evolution. Our ability to communicate clearly and accurately continues to be the most powerful adaptation in our toolkit. For this reason, developing a simple, clear, and comprehensible lexicon is a vital part of this project, which is why the final part of this book has focused on precisely this task. In the end, we have proposed *evidist* and *evidism* as unifying terms to transcend tribe, religion, and any set of values, denoting any person who recognizes the profound importance of the scientific way of knowing. We hope our argument for these terms has been convincing. But even if you remain unconvinced about the terms themselves, we hope you are convinced of the importance of the concept they represent— the embrace, by as many people as possible, of the scientific way of knowing.

If you are convinced of this, we have a final request: do not keep it to yourself.

Knowing that, from the perspective of Earth, the surface at the equator is moving at 1000 miles per hour, or that water contains two hydrogen atoms and one oxygen atom, or that time moves more slowly as you move more quickly—these are the kinds of scientific facts that can stay locked inside your mind if you wish, to be turned over and examined in solitude. But understanding that the scientific way of knowing is the best way of uncovering reality, the best hope for solving our most urgent problems, and the best way of reducing conflict for mankind—this you must not keep to yourself. Use the concepts and techniques described in this book to bring those you know into the community of evidists.

Help them join the evolution, for everyone's sake.

# APPENDIX

# COMPLETE LIST
# OF SUGGESTIONS

1. Explain that traditional ways of knowing, including religions, are natural and often wrong.

2. Teach and demonstrate the three minimal elements of scientific understanding: (1) science is reliable, (2) anyone can look up what is known, and (3) nothing is known for certain.

3. Teach people to separate facts from non-facts.

4. Accepting the truth about all facts is important—mere critical thinking, secularism, skepticism, and atheism are not enough.

5. A false understanding of facts impedes development of good values.

6. Be cocksure about established facts and cautious about values.

7. Be wary of values that perpetuate unworthy aspects of human nature.

8. Teach that natural human tendencies toward tribalism should be transcended or channeled appropriately.

9. Urge each member of a religious congregation to accept the scientific consensus on facts.

10. Use the word *religion* in ways that allow consistency with scientific consensus on the facts.

11. Show that you are willing to state publicly your acceptance of science and the importance of religions changing to be consistent with the facts.

12. Explain the correct meanings of the words *agnostic* and *atheist*.

13. Replace most uses of the word "atheism" with "antitheism" or "humanism" or "evidism."

14. Be alert to other meanings of the word secular.

15. Point out subtle supernatural and tribalistic messages in mass media.

16. Encourage education.

17. Use the word *secular* in a way that allows people who call themselves religious to also be *secular*.

18. When advocating secularism, avoid arguing for irrelevant values on which people disagree.

19. When marketing the best way of knowing facts, use an effective label.

20. Adopt a marketable label for people who accept the scientific way of knowing.

21. Update the commonly used definitions of *religion*.

22. Adopt an effective label for the traditional ways of knowing.

23. To achieve good education in public schools, be politically active.

24. Encourage a community to make the scientific way of knowing their norm.

25. Urge legislators to exclude from public benefit programs organizations that promote science denial.

# ACKNOWLEDGMENTS

Richard Carrier, Russell Blackford, and Shadia Drury gave helpful substantive critiques of the entire book. Johnny Monsarrat contributed portions of chapter 2. William R. Murry contributed information for chapter 3. Ron Lindsay critiqued our explanation of secularism. Mynga Futrell helped us find the word "evidist." Ed Buckner, Tom Flynn, Austin Dacey, and Alex Rosenberg offered helpful comments on an early draft.

# ABOUT THE AUTHORS

**Jeff T. Haley** is an inventor (patents 7,366,913, 8,349,120, 8,442,490, 8,865,133, 9,100,824, EU 2037941), entrepreneur (OraHealth), chemist, clinical researcher, patent lawyer, public interest advocate (founder, co-sponsor, chair of Washington's successful medical marijuana initiative campaign), and former civil rights lawyer (Bethel v. Fraser, USSC). He can be reached at jeff@haley.net.

**Dale McGowan** is an author/editor of books, including *Parenting Beyond Belief, Raising Freethinkers, Atheism for Dummies,* and *In Faith and in Doubt.* He was the founding executive director of Foundation Beyond Belief, a humanist nonprofit that raises funds for charities worldwide, promotes volunteerism, and organizes the humanist response to disaster recovery and international service. He can be reached at dale@dalemcgowan.com.